the Super Source™

Geometry

Cuisenaire Company of America, Inc.
Orangeburg, NY

Cuisenaire extends its warmest thanks to the many teachers and students across the country who helped ensure the success of the Super Source™ series by participating in the outlining, writing, and field testing of the materials.

Managing Editor: Alan MacDonell
Developmental Editor: Deborah J. Slade
Contributing Writer: Carol Desoe

Design Manager: Phyllis Aycock
Cover Design: Phyllis Aycock
Text Design: Fiona Santoianni
Line Art and Production: Fiona Santoianni

Printed in the United States of America
ISBN 1-57452-011-3

1 2 3 4 5 - SG - 02 01 00 99 98

the Super Source™

Table of Contents

Using the Super Source™

The Super Source™ Grades 7-8 continues the Grades K-6 series of activities using manipulatives. Driving ***the Super Source*** is Cuisenaire's conviction that children construct their own understandings through rich, hands-on mathematical experiences. There is a substantial history of manipulative use in the primary grades, but it is only in the past ten years that educators have come to agree that manipulatives play an important part in middle grade learning as well.

Unlike the K-6 series, in which each book is dedicated to a particular manipulative, the Grades 7-8 series is organized according to five curriculum strands: Number, Geometry, Measurement, Patterns/Functions, and Probability/Statistics. The series includes a separate book for each strand, each book containing activities in which students use a variety of manipulatives: Pattern Blocks, Geoboards, Cuisenaire® Rods, Snap™ Cubes, Color Tiles, and Tangrams.

Each book contains eighteen lessons grouped into clusters of 3-5 lessons each. Each cluster of lessons is introduced by a page of comments about how and when the activities within each lesson might be integrated into the curriculum.

The lessons in ***the Super Source*** follow a basic structure consistent with the vision of mathematics teaching described in the *Curriculum and Evaluation Standards for School Mathematics* published by the National Council of Teachers of Mathematics. All of the activities involve Problem Solving, Communication, Reasoning, and Mathematical Connections—the first four NCTM Standards.

HOW THE LESSONS ARE ORGANIZED

At the beginning of each lesson, you will find, to the right of the title, a list of the topics that students will be working with. GETTING READY, offers an *Overview*, which states what children will be doing, and why, and provides a list of *What You'll Need.* Specific numbers of manipulative pieces are suggested on this list but can be adjusted as the needs of your particular situation dictate. In preparation for an activity, pieces can be counted out and placed in containers or self-sealing plastic bags for easy distribution. Blackline masters that are provided for your convenience at the back of the book are also referenced on this materials list, as are activity masters for each lesson.

The second section, THE ACTIVITY, contains the heart of the lesson: a two-part *On Their Own*, in which rich problems stimulate many different problem-solving approaches and lead to a variety of solutions. These hands-on explorations have the potential for bringing students to new mathematical ideas and deepening skills. They are intended as stand-alone activities for students to explore with a partner or in a small group. *On Their Own* Part 2 extends the mathematical ideas explored in Part 1.

After each part of *On Their Own* is a *Thinking and Sharing* section that includes prompts you can use to promote discussion. These are questions that encourage students to describe what they notice, tell how they found their results, and give the reasoning behind their conclusions. In this way, students learn to verify their own results rather than relying on the teacher to determine if an answer is "right" or "wrong." When students compare and discuss results, they are often able to refine their thinking and confirm ideas that were only speculations during their work on the *On Their Own* activities.

The last part of THE ACTIVITY is *For Their Portfolio*, an opportunity for the individual student to put together what he or she has learned from the activity and discussion. This might be a piece of writing in which the student communicates results to a third person; it could be a drawing or plan synthesizing what has occurred; or it could be a paragraph in which the student applies the ideas from the activity to another area. In any case, the work students produce *For Their Portfolio* is a reflection of what they've taken from the activity and made their own.

The third and final section of the lesson is TEACHER TALK. Here, in *Where's the Mathematics?,* you as the teacher can gain insight into the mathematics underlying the activity and discover some of the strategies students are apt to use as they work. Solutions are also given, when such are necessary and/or helpful. This section will be especially helpful to you in focusing the *Thinking and Sharing* discussion.

USING THE ACTIVITIES

The Super Source is designed to fit into a variety of classroom environments. These can range from a completely manipulative-based classroom to one in which manipulatives are just beginning to play a part. You may choose to have the whole class work on one particular activity, or you may want to set different groups to work on two or three related activities. This latter approach does not require full classroom sets of a particular manipulative.

If students are comfortable working independently, you might want to set up a "menu"—that is, set out a number of activities from which students can choose. If this is the route you take, you may find it easiest to group the lessons as they are organized in the book—in small clusters of related activities that stimulate similar questions and discussion.

However you choose to use ***the Super Source*** activities, it would be wise to allow time for several groups or the entire class to share their experiences. The dynamics of this type of interaction, where students share not only solutions and strategies but also thoughts and intuitions, is the basis of continued mathematical growth. It allows students who are beginning to form a mathematical structure to clarify it and those who have mastered isolated concepts to begin to see how these concepts might fit together.

At times you may find yourself tempted to introduce an activity by giving examples or modeling how the activity might be accomplished. Avoid this. If you do this, you rob students of the chance to approach the activity in their own individual way. Instead of making assumptions about what students will or won't do, watch and listen. The excitement and challenge of the activity—as well as the chance to work cooperatively—may bring out abilities in students that will surprise you.

USING THE MANIPULATIVES

Each activity in this book was written with a specific manipulative in mind. The six manipulatives used are: Geoboards, Color Tiles, Snap Cubes, Cuisenaire Rods, Pattern Blocks, and Tangrams. All of these are pictured on the cover of this book. If you are missing a specific manipulative, you may still be able to use the activity by substituting a different manipulative. For example, most Snap Cube activities can be performed with other connecting cubes. In fact, if the activity involves using the cubes as counters, you may be able to substitute a whole variety of manipulatives.

The use of manipulatives provides a perfect opportunity for authentic assessment. Watching how students work with the individual pieces can give you a sense of how they approach a mathematical problem. Their thinking can be "seen" in the way they use and arrange the manipulatives. When a class breaks up into small working groups, you can circulate, listen, and raise questions, all the while focusing on how your students are thinking.

Work with manipulatives often elicits unexpected abilities from students whose performance in more symbolic, number-oriented tasks may be weak. On the other hand, some students with good memories for numerical relationships have difficulty with spatial reasoning and can more readily learn from free exploration with hands-on materials. For all students, manipulatives provide concrete ways to tackle mathematical challenges and bring meaning to abstract ideas.

Overview of the Lessons

Geometry, Grades 7-8

Investigating Congruence and Similarity

Each of the lessons in this cluster can be used to introduce new material or to reinforce ideas that have already been explored in class. These ideas include congruence, reflections, rotations, ratio, similarity, congruent corresponding angles in similar figures, proportional corresponding sides in similar figures.

1. Cardboard Cartons (*Exploring congruence*)

This activity allows students to explore congruence, reflection, and rotation. It could be used as an introduction to the topics, but because *On Their Own* Part 1 uses all three terms, teachers should at least give a brief explanation of those terms before starting the activity. This might be readily accomplished by drawing the figures shown below on the board (without the labels) and discussing which are/are not congruent and why.

S S — Congruent

s **S** — Not congruent (different sizes)

S Ƨ — Congruent through reflection

S ᔕ — Congruent through rotation

Before starting the activity, teachers should also clarify what is and is not a hexomino. (A clear explanation is given for the teacher at the beginning of *Where's the Mathematics*, page 11.)

The lesson is written as a Color Tiles activity, but could be done with Snap™ Cubes instead.

2. Gulliver's Shapes (*Exploring similarity*)

This activity gives an informal introduction to similarity, stressing shape and size comparisons without going into depth on formal definitions of similarity. The term "similarity" is not used in *On Their Own* Part 1, but it is used in Part 2. Therefore, the discussion following Part 1 would be a good place to introduce the term (or ask for it), if it is unfamiliar to some students. Part 2 uses the terms "corresponding angle measures" and "corresponding side lengths" but does not generalize about them.

Part 2 suggests (but doesn't require) using a protractor.

For Their Portfolio requires an understanding of the term "scale factor." This could be introduced in the discussion(s) following Part 1 and/or Part 2, perhaps with a question like "What do you have to multiply this side length by to get the corresponding length on the other figure? That's the scale factor."

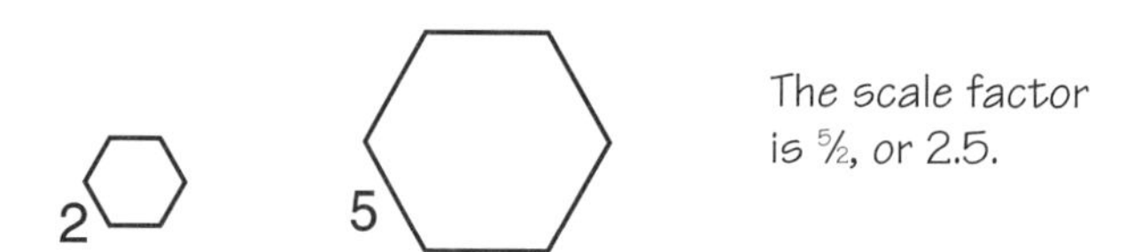

The scale factor is $\frac{5}{2}$, or 2.5.

3. Tan's House (*Extending similarity*)

This activity can be used to reinforce intuitive notions of similarity and introduce formal aspects of similarity.

On Their Own Part 1 assumes that students understand that similar figures are figures with the same shape, regardless of size. It also assumes that students understand the term "ratio."

The activity allows students to explore the concepts of proportional corresponding sides and congruent corresponding angles between similar figures.

The activity provides an opportunity (at the teacher's discretion) to use the Pythagorean Theorem. (See *Where's the Mathematics?*, page 20.)

CARDBOARD CARTONS

- Spatial visualization
- Congruence
- Transformations

Getting Ready

What You'll Need

Color Tiles, at least 60 per pair

1-inch grid paper, page 106

Scissors

Tape

Activity Master, page 88

Overview

Students search to find all possible arrangements of six Color Tiles. They then determine which of these arrangements could be folded to form a cube. In this activity, students have the opportunity to:

- devise strategies for finding different arrangements
- reinforce understanding of the concept of congruence
- strengthen spatial visualization skills

Other *Super Source* activities that explore these and related concepts are:

Gulliver's Shapes, page 13

Tan's House, page 17

The Activity

On Their Own (Part 1)

Tino has a large sheet of cardboard and 6 square tiles that measure 12 inches on each edge. How many different 6-tile arrangements (hexominoes) can Tino trace on the cardboard sheet if at least one complete edge of each tile must touch one complete edge of another tile?

- Working with a partner, use 6 Color Tiles to make as many hexominoes as you can.
- Record your models on 1-inch grid paper, cut them out, and decide on a way to sort them.
- Make sure that each hexomino is different from the others. Eliminate hexominoes that are congruent to others through reflections (flips) and/or rotations.
- Exchange hexominoes with another group. Check to see that none of their hexominoes are congruent. Mark any that you think are the same. Be ready to justify your findings.
- Return the hexominoes. Check yours to see if any duplicates were found.

Thinking and Sharing

Invite one or two pairs of students to present their hexominoes in an organized way. Ask students to identify congruent hexominoes, missing hexominoes, and figures that are not hexominoes.

Use prompts like these to promote class discussion:

- How did you go about searching for hexominoes?
- Did you use a strategy to find new shapes? If so, explain what you did.
- Did you find any patterns while making your hexominoes? How did you use these patterns to help you sort the hexominoes?
- Did sorting your hexominoes help you find others that were missing? If so, explain.
- How did you use reflections and/or rotations of hexominoes to discover congruent shapes?

On Their Own (Part 2)

What if... ***Tino wants to send his friend a new basketball for his birthday? The basketball will be packed in a cubic cardboard carton that measures 12 inches on each edge. How could Tino construct the carton from the sheet of cardboard?***

- Examine your hexominoes and predict which of them could be folded along the lines to form a cubic cardboard carton.
- Build models of your selections. You may use whatever materials you have available.
- If packing tape is used to secure the unfolded edges of the carton as it is assembled, decide whether any of your models require less tape than others. Be ready to explain your reasoning.

Thinking and Sharing

Have students display their findings and explain what they looked for when searching for hexominoes that would fold into cartons.

Use prompts like these to promote class discussion:

- How did you go about figuring out which hexominoes could be folded to form cubic cartons?
- How many faces, edges, and vertices do your cubic cartons contain?
- How can the number of fold lines on the hexomino be used to determine which models can be folded to form a cubic carton?
- How can you predict the number of edges that must be taped before the carton is assembled?

Write a short paragraph describing the differences between hexominoes that will form cubic cartons and those that won't. Include any diagrams that might help illustrate your points.

Teacher Talk

Where's the Mathematics?

Students will use a wide variety of strategies to search for all possible hexominoes. Be sure they understand how the tiles must be adjoined. Point out that the tiles must be laid edge-to-edge as shown below.

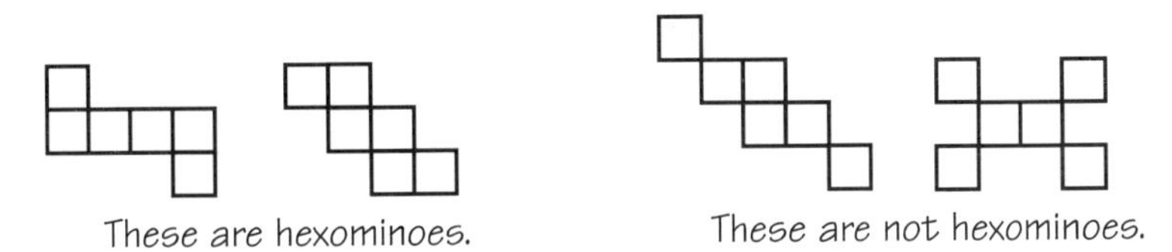

Students may be surprised at the number of different hexominoes they are able to generate. It should not be expected that all students will find all the different hexominoes. However, in Part 1, each pair of students should find a significant number and be able to eliminate duplicates from their collection. Some students may decide to make as many hexominoes as they can and then check for duplicates when they think they have created every possible hexomino. Other students may check for duplications as they go along. The 35 hexominoes are shown below.

Using their cutout models of the hexominoes, students should recognize that if one hexomino can be reflected (flipped) and/or rotated to match another hexomino exactly, the shapes are congruent and, therefore, are not different hexominoes. These duplicates should be eliminated.

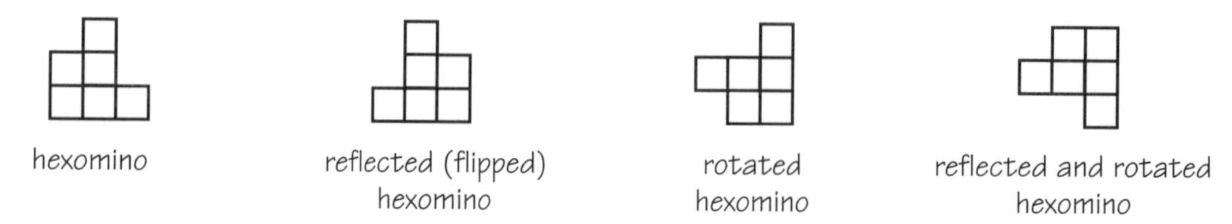

In Part 2, many students will find several hexominoes that can be folded into the cubic carton. Students will discover, however, that all of these cartons have exactly 12 edges, 8 vertices, and 6 faces. By counting the number of fold lines on the hexomino models, students should be able to predict the number of edges that will need to be taped when the cartons are assembled. They may be surprised to find that no matter which model they choose, there will be five folded edges and seven edges that require taping. Hexominoes that will form cubic cartons are shown below.

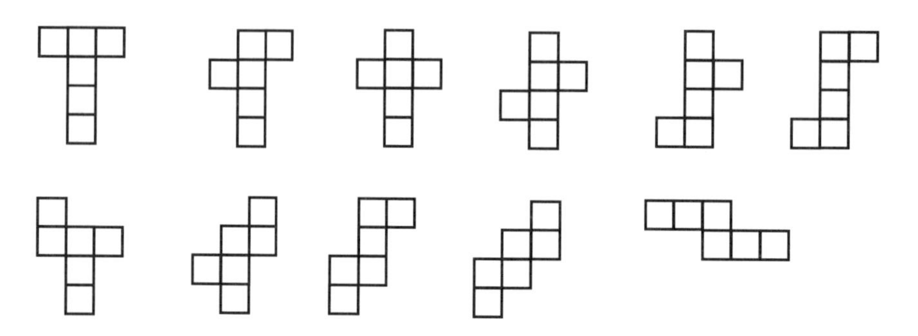

Students will have different strategies for building, checking, and sorting their models. By sharing their strategies, students can learn about a variety of problem-solving approaches that can be used to solve these and other types of problems.

GULLIVER'S SHAPES

- Similarity
- Spatial visualization
- Pattern recognition

Getting Ready

What You'll Need

Pattern Blocks, 1 set per group

Scissors

Rulers

Protractors

Activity Master, page 89

Overview

Students create enlargements of individual Pattern Block shapes and reductions of their enlargements. In this activity, students have the opportunity to:

- devise strategies to prove two shapes are similar
- reinforce understanding of the concept of similarity
- recognize patterns useful in making predictions and solving problems
- strengthen spatial visualization skills

Other *Super Source* activities that explore these and related concepts are:

Cardboard Cartons, page 9

Tan's House, page 17

The Activity

On Their Own (Part 1)

In Jonathan Swift's novel Gulliver's Travels, Gulliver finds himself in the great city of Lilliput. Here everything, including its citizens, the Lilliputians, are very small in comparison to Gulliver and his belongings. If the Lilliputians had a set of Pattern Blocks in their schoolroom, Gulliver would find them much too small for his use. Can you build enlargements of each of the Pattern Block shapes that Gulliver could use?

- Working with your group, try to build at least 4 different-sized enlargements of each shape in the Pattern Block set. For each enlargement, use only blocks that are congruent to the original shape, and only one layer of blocks.
- Record the number of blocks used to build each enlargement. Look for patterns as you work. If you are unable to build an enlargement of a shape, try to figure out why.
- Trace only the outlines of each enlargement onto unlined paper and cut them out.
- Compare the angle measures of each enlargement to those of the original Pattern Block shape. Compare the lengths of the sides of each enlarged shape to those of the original shape.
- Summarize the findings of your group.

Thinking and Sharing

Ask students to explain how they built enlargements of the original Pattern Block shapes. Have them compare the number of blocks used in each enlargement and discuss any patterns they noticed that may have helped them in making the similar shapes.

Use prompts like these to promote class discussion:

- For which blocks were you able to make enlargements that were similar to the original block?
- Which block was the hardest to enlarge? Why?
- For which block was it impossible to make a larger similar block? Why do you think it was impossible?
- How did you determine if each new shape you made was similar to the original shape?
- What pattern(s), if any, did you discover in the number of blocks used to build the set of enlargements?
- How did this pattern help you build other similar figures?

On Their Own (Part 2)

What if... ***Gulliver found himself in Brobdingnag, an imaginary land of giants, where students used Pattern Block shapes that were much too large for him to work with? Can you create reductions of each of the enlarged Pattern Block shapes that Gulliver could use?***

- Exchange the cutouts of the biggest enlargement for each Pattern Block shape with those of another group. Imagine these enlargements are congruent to the Pattern Blocks used in Brobdingnag.
- Working from your cutouts, try to build at least 3 different scaled-down, similar versions of each Brobdingnag block. You may use a protractor and ruler and/or paper folding techniques on the cutouts, but you may not use Pattern Blocks or your other enlargements.
- Trace the outlines of these smaller shapes onto unlined paper.
- Compare your reductions to the original cutouts by considering corresponding angle measures and corresponding side lengths.
- Be ready to explain your methods.

Thinking and Sharing

Invite students to talk about how they created their scaled-down shapes. If other groups have different approaches, invite them to share their work.

Use prompts like these to promote class discussion:

- What strategies did you use to generate each of the scaled-down shapes?
- How did you determine if each new shape you made was similar to the original shape?

- Which shape was the hardest to reduce? Why?
- How many enlargements and reductions do you think there are for each Pattern Block?
- What is the relationship between the reductions and enlargements of each Pattern Block shape?

Suppose a friend wants to come to your house after school to visit. Using your knowledge of similarity, draw a scale map of the route he must follow from the school to your house. Be sure to include the scale factor, landmarks, and street names.

Teacher Talk

Where's the Mathematics?

Students are likely to go through numerous trial-and-error processes as they begin the activity. Through their investigations, they should discover that it is possible to build enlargements of all of the Pattern Blocks except the hexagon (although enlargements of the hexagon can be built using combinations of different Pattern Blocks). In Part 1, as students arrange their enlargements in order of increasing size, they may notice a pattern in the number of blocks needed to build each similar shape. The number of blocks needed follows the sequence of perfect squares: 1, 4, 9, 16, 25, and so on. Students can use this sequence to figure out the number of blocks needed for each future enlargement. This may make the task easier, but will not eliminate trial and error.

Enlargements of three of the Pattern Blocks are shown below.

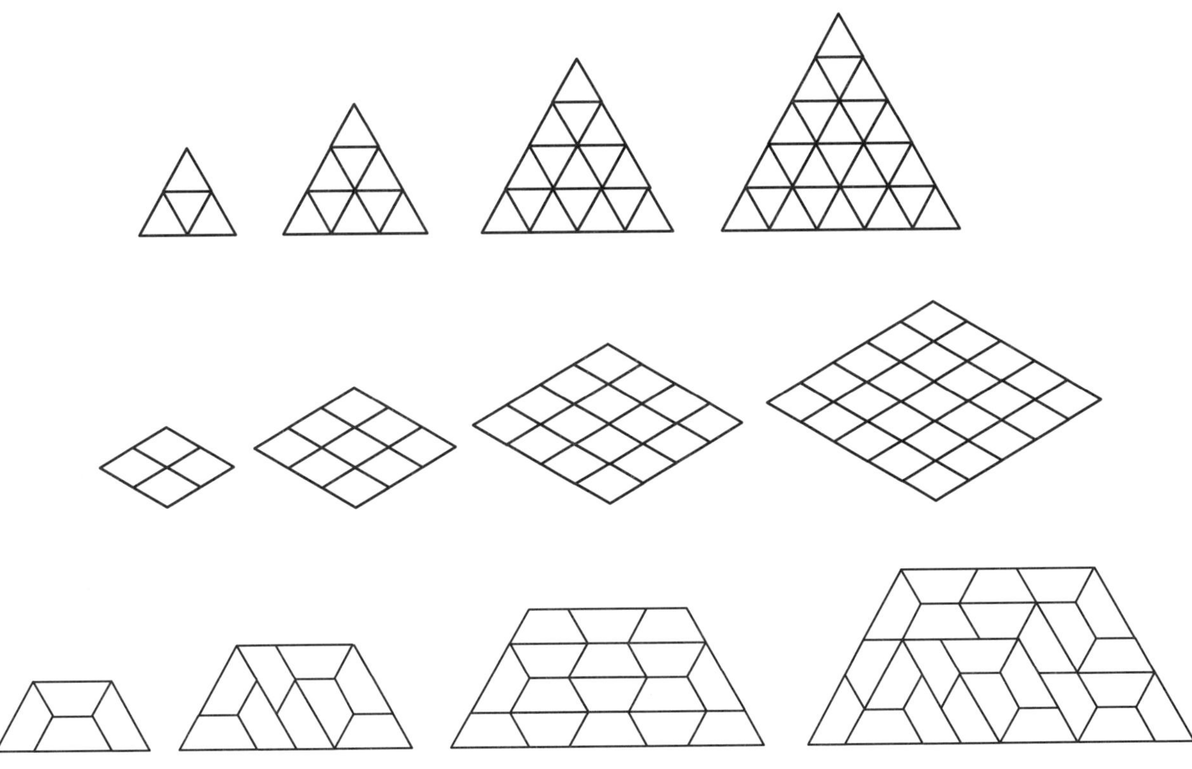

Students will find it impossible to build larger hexagons similar to the original shape using only hexagon blocks because their 120° angles cannot be combined in any way to form a straight angle (180°) at any of the existing vertices. It is possible, however, to build similar hexagons using combinations of different Pattern Blocks.

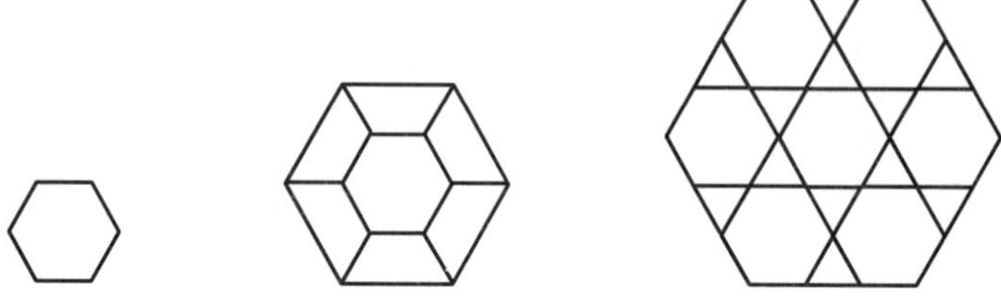

The second hexagon in the above series has side lengths that are twice the length of those in the original hexagon. Because each trapezoid has half the area of a hexagon, this figure has an area equivalent to that of 4 hexagons $(6(\frac{1}{2}) + 1 = 4)$. The third hexagon has side lengths that are three times those in the original hexagon. Because each triangle has one sixth the area of a hexagon, this figure has an area equivalent to that of 9 hexagons $(12(\frac{1}{6}) + 7 = 9)$.

Students should observe that corresponding angles of similar shapes are congruent. They may also recognize that corresponding sides are proportional. Some students may overlay their scale drawings on top of each other to compare them. Others may use their blocks to measure and compare the angles and sides of their enlarged shapes.

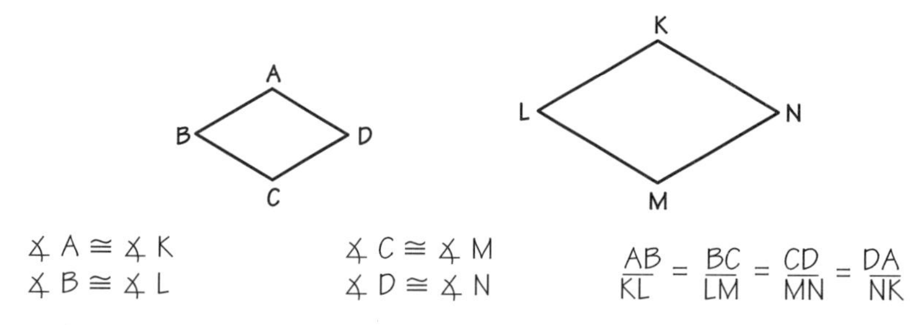

In Part 2, some students may use a method that combines tracing and measuring (using a protractor and a ruler) to create smaller similar shapes with congruent angles and proportional sides. Others may experiment with paper-folding. To create similar shapes using paper-folding, each of the folds used to make the reduced shape needs to be parallel to one of the original sides, thus ensuring that the newly-formed angles are congruent to those of the original shape. The diagrams below show ways to use paper-folding to create reductions of the square and equilateral triangle.

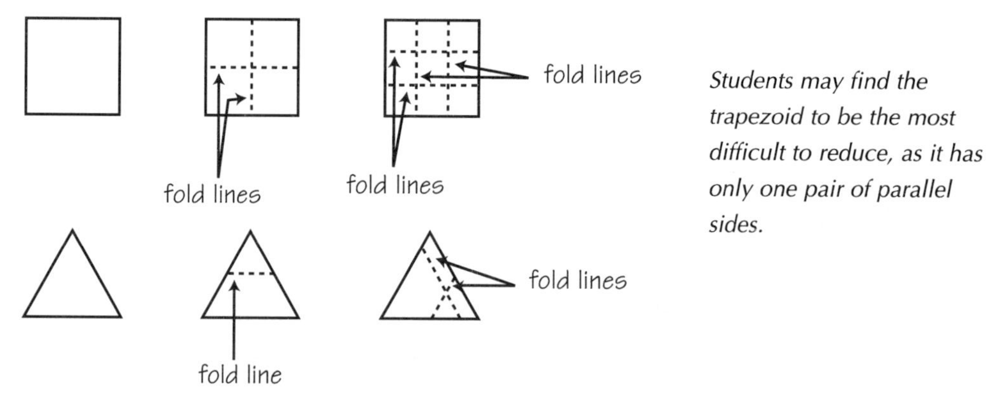

Students may find the trapezoid to be the most difficult to reduce, as it has only one pair of parallel sides.

Students should come to realize that there are infinitely many enlargements and/or reductions that can be made of each Pattern Block. They should also recognize that the corresponding angles of the enlargements and reductions are congruent, and that the lengths of their corresponding sides are in proportion. Their recognition of these relationships will help students to better understand the concept of similarity.

TAN'S HOUSE

- Similarity
- Ratio
- Proportion

Getting Ready

What You'll Need

Tangrams, 1 set per pair

Tangram paper, page 107

Rulers

Protractors

Scissors

Activity Master, page 90

Overview

Students construct two sets of Tangrams in which the side lengths of the pieces of each new set are multiples of the lengths of existing pieces. In this activity, students have the opportunity to:

- apply the concept of similarity to pairs of polygons
- use ratios and proportions to measure and enlarge polygons
- draw scale models

Other *Super Source* activities that explore these and related concepts are:

Cardboard Cartons, page 9

Gulliver's Shapes, page 13

The Activity

On Their Own (Part 1)

Tan is an architect who builds houses like the one shown based on Tangram shapes. He needs to construct a set of Tangram pieces whose sides measure three times the lengths of the sides of those in his original set. How might he do this?

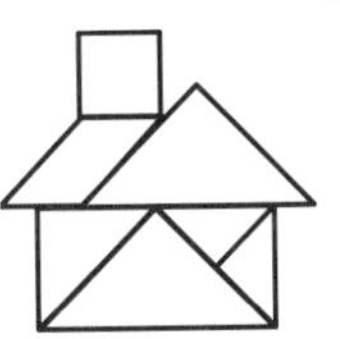

- Working with a partner, use your Tangram set as a blueprint to construct a larger set of Tangram pieces. Make the ratio of the side lengths of each plastic Tangram piece to the side lengths of the corresponding larger piece 1:3.
- Copy the new pieces onto Tangram paper and cut them out. Call this new set *Tangram 2.*
- Be prepared to explain how you constructed each piece, and how you know each piece is similar to the original Tangram piece.

Thinking and Sharing

Have students compare the shapes they made with those made by other pairs. Encourage them to share the methods they used to construct their shapes.

Use prompts like these to promote class discussion:

- What strategies did you use to make the enlargements of the pieces? Did you use the same strategies for all of the pieces? If not, why not?

- What relationships exist between the angles of the enlarged pieces and those of the original pieces? What relationships exist between the sides of the enlarged pieces and those of the original pieces?
- Which piece was the easiest to enlarge? Which piece was the hardest? Why?
- Did you have to measure every side of each piece before enlarging it? Why or why not?
- What pieces from the Tangram set were already similar? How did you know?

On Their Own (2)

What if... ***Tan decides the pieces in Tangram 2 are not large enough? What if he wants a third set of Tangrams where the medium triangle from Tangram 2 becomes the small triangle in this new set?***

- Using the original plastic Tangram set and/or your new set, *Tangram 2*, as a blueprint, construct a larger set of Tangram pieces in which the medium triangle of *Tangram 2* becomes the small triangle of the third set.
- Copy the new set of pieces onto Tangram paper and cut them out. Call this new set *Tangram 3*.
- Decide whether or not each new piece is similar to the piece in the original plastic Tangram set. Justify your answer.

Thinking and Sharing

Have students compare their *Tangram 3* set with those made by other groups. Encourage them to share the methods they used to construct this set of shapes.

Use prompts like these to promote class discussion:

- What strategies did you use to make the enlargements of the pieces? Did you use the same strategies for all of the pieces? If not, why not?
- Was it easier to use the original set or to use the *Tangram 2* set as a blueprint? Did anyone use a combination of sets? If so, tell what you did.
- Which piece was the easiest to enlarge? Which piece was the hardest? Why?
- What relationships exist among the three sets of Tangram pieces? How do you know?
- What are the proportional relationships between the side lengths of the smallest triangles in the three sets of Tangrams? How does this information help in determining the proportional relationships of the three Tangram sets?

Using all of the pieces in the original Tangram set, design a house that is different from Tan's house. Trace its blueprint on paper, indicating where each shape is located. Then, using either *Tangram 2* or *Tangram 3*, build a similar enlargement of the original house and trace its blueprint on paper, indicating the position of its shapes.

Teacher Talk

Where's the Mathematics?

As students work through these activities, they observe how the side lengths and angle measurements of similar polygons are related. In addition, they learn to use problem-solving and measuring skills, while applying equalities, ratios, and proportions, in a concrete situation.

Within the original set of Tangram pieces, students may notice that the five triangles are similar to each other. They may recognize that their corresponding angles are congruent and that their corresponding sides are in proportion. In the same manner, students may also discover that corresponding angle measures of the enlarged shapes in the Tangram sets they make are equal to those in the original set. A comparison of the sides of two similar pieces will reveal that the ratios of the lengths of corresponding sides are all equal.

Students may tackle the problem of making enlargements of the Tangram pieces in a variety of ways. In Part 1, some students may triple each side of the original square built from one set of Tangrams. Students may then choose to fold or draw lines within the square to make the larger Tangram pieces.

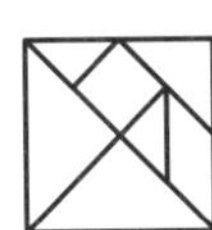

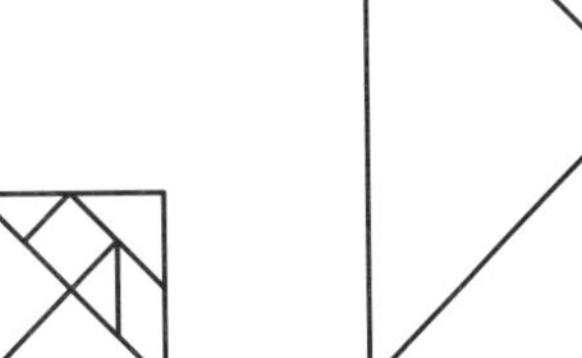

Other students may choose to enlarge pieces of the Tangram set one at a time. They may use a ruler and a protractor or the Tangram pieces themselves to measure the sides and angles.

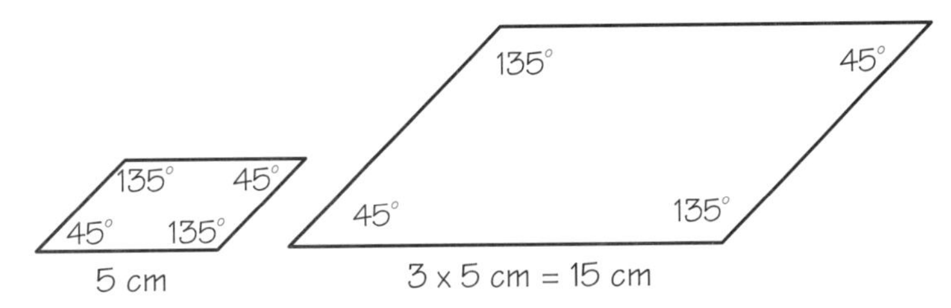

Students may realize that the small Tangram triangle can be used to generate the other Tangram pieces in the set. This method can also be used to generate the third set of Tangram pieces in Part 2. When this method is used, the remaining triangle, square, and parallelogram shapes are all easy to construct.

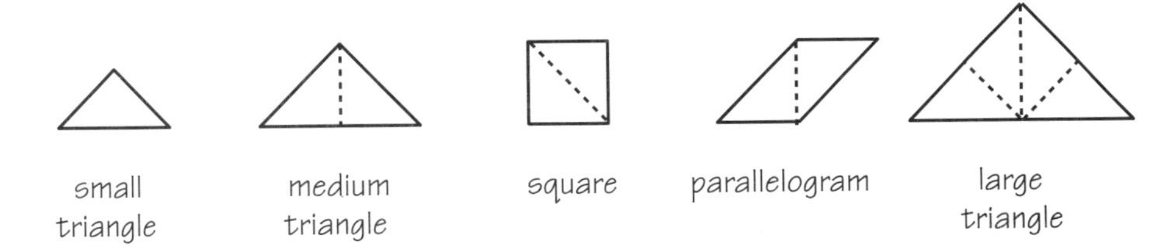

Because the same ratio exists between the lengths of the sides of corresponding pieces in any two Tangram sets, students can focus on the small triangle in each set to discover the proportions between the original Tangram set, *Tangram 2,* and *Tangram 3*. The Pythagorean Theorem, $a^2 + b^2 = c^2$, can be helpful in finding the lengths of the sides of the enlarged triangles, as the Tangram triangles are all right triangles.

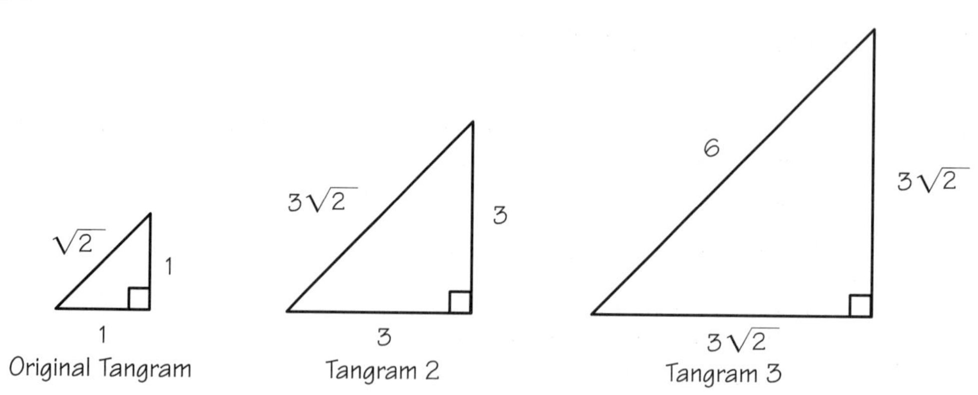

Therefore, the ratio of the side lengths of the pieces in the original Tangram set to those in *Tangram 2* to those in *Tangram 3* is $1 : 3 : 3\sqrt{2}$.

Investigating Triangles

Each of the lessons in this cluster can be used to introduce new material or to reinforce ideas that have already been explored in class. These ideas include classification, area, and perimeter of triangles; the Triangle Inequality Theorem; theorems involving inscribed and central angles; and properties of isosceles and right triangles.

1. Rooftop Triangles (*Exploring the Triangle Inequality Theorem*)

This activity provides a discovery approach to the Triangle Inequality Theorem. In *On Their Own* Part 1 students try to make triangles with sides of different lengths and form conclusions about which will work and which won't. The discussion questions categorize triangles as equilateral, scalene, and isosceles; but these terms are not used in the *On Their Own* and can be introduced here by the teacher or else serve as reinforcement for students who already know the terms.

2. Shelf Brackets (*Investigating the area of right triangles*)

This activity asks students to find the areas of various right triangles but does not prescribe a particular method. The geoboard makes it possible for students to calculate areas by adding partial square units, by using the standard area formula, or by using a method based on the Area Addition Theorem, described in *Where's the Mathematics* (page 29).

Both *On Their Owns* assume that students can recognize right triangles. *On Their Own* Part 2 also assumes an intuitive knowledge of the terms "reflect" and "rotate."

3. Hydroponics (*Exploring isosceles triangles and central and inscribed angles*)

In this activity students discover that base angles of isosceles triangles are congruent. The activity also uses isosceles triangles as a forum for exploring theorems about the intercepted arcs of central and inscribed angles.

On Their Own Part 1 provides a definition of "isosceles" and spells out two theorems about intercepted arcs of central and inscribed angles. This makes the activity usable either as an introduction to or as reinforcement of this material.

Where's the Mathematics? suggests (top of page 33) how teachers might introduce the idea that congruent chords subtend congruent arcs.

4. Sal's Similar Sails (*Investigating area and perimeter of triangles*)

This activity uses nonstandard measure to calculate the area and perimeter of triangles. Students discover that the triangles they can build are all similar. The activity is a good way to reinforce similarity topics, including proportionality of sides and congruence of angles.

On Their Own Part 1 assumes an understanding of area, and *On Their Own Part* 2 assumes an understanding of area and perimeter.

Where's the Mathematics? contains suggestions for extending the topic to the Pythagorean Theorem, irrational numbers, and decimal approximations using calculators. For teachers who want to introduce more advanced topics, it also illustrates the idea that the ratio of areas of similar triangles is the square of the ratio of corresponding sides.

ROOFTOP TRIANGLES

- Properties of triangles
- Triangle inequality theorem
- Organizing data

Getting Ready

What You'll Need

Cuisenaire Rods, 1 set per pair

Activity Master, page 91

Overview

Students investigate combinations of three Cuisenaire Rods that can be placed corner-to-corner to form triangles. In this activity, students have the opportunity to:

- explore the relationships that exist among the lengths of the sides of a triangle
- organize, record, and analyze data
- discover the Triangle Inequality Theorem

Other *Super Source* activities that explore these and related concepts are:

Shelf Brackets, page 26

Hydroponics, page 30

Sal's Similar Sails, page 34

The Activity

On Their Own (Part 1)

A lumber yard stocks wooden beams used to make triangular trusses for roof supports. They stock beams in lengths of 1 yard, 2 yards, 3 yards, 4 yards, and so on, up to 10 yards. What combinations of lengths can be used to make the triangular trusses?

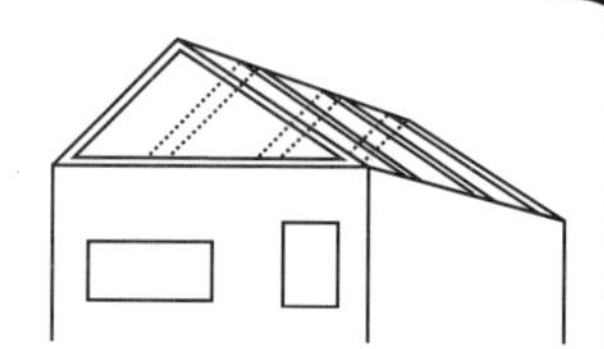

- Working with a partner, use Cuisenaire Rods to build models of triangular trusses. Arrange the rods so that each rod touches a corner of the other rods.

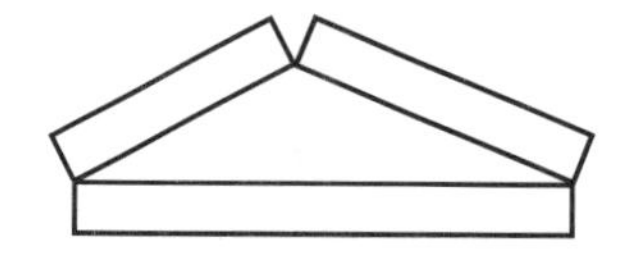

- Try at least 15 different 3-rod combinations. Record the combinations of lengths that form triangles in one list and the combinations that do not form triangles in another list.
- Organize and analyze your data. Try to figure out why some 3-rod combinations will form triangles while others will not. Be ready to explain your reasoning.

Thinking and Sharing

Have students help you create a class chart listing the combinations that formed triangles in one column, and those that did not in the other.

Use prompts like these to promote class discussion:

- What patterns did you notice?
- How does a combination of rods that forms a triangle differ from a combination that does not?
- Will rods with lengths 2 cm (red), 3 cm (light green), and 8 cm (brown) form a triangle? Why or why not?
- If you begin to build a triangle with a 3-cm rod (light green) and a 5-cm rod (yellow), which length or lengths could you use to complete the triangle? Explain your thinking.
- How would you describe the relationship that exists between the lengths of rods that form triangles? Support your conclusion with examples.

On Their Own (Part 2)

What if... ***the lumber company wants to list in its merchandise catalog all possible triangular roof trusses it can supply from its stock? How can they organize their list to be sure to include all possible combinations?***

- Working with your partner and another pair of partners, find all possible trusses the lumber company can build. Use your Cuisenaire Rods to model the trusses and to check your combinations.
- Organize your data in a way that you think would be practical for the catalog listing.

Thinking and Sharing

Have groups compare their data and describe how they organized their lists.

Use prompts like these to promote class discussion:

- How did you organize your search for possible combinations?
- How did you sort and organize your data?
- How many of the combinations form equilateral triangles? isosceles triangles? scalene triangles?
- Are there some combinations that you think are more (or less) likely to be ordered by builders than others? If so, which ones, and why?

Note that some students may consider a 3-4-5 truss different from a 4-3-5 truss or a 5-4-3 truss if they are envisioning, for example, the first length as the base of the roof support. They may reason that the roof will take on different shapes depending upon which length forms the base, and, thus, the trusses are different.

Write a letter to the owner of the lumber yard explaining how to determine possible lengths for the third side of a triangular roof support (without physically trying them) if the lengths of two sides of the triangular support are known.

Teacher Talk

Where's the Mathematics?

Students' work on this activity should lead them to see that not all combinations of three lengths can be joined to form a triangle. In fact, the only combinations that will work are ones in which the sum of any two of the lengths is greater than the third length (the Triangle Inequality Theorem). If the sum of the lengths of the two shorter sides of the triangle is less than or equal to the length of the remaining side, a triangle will not be formed.

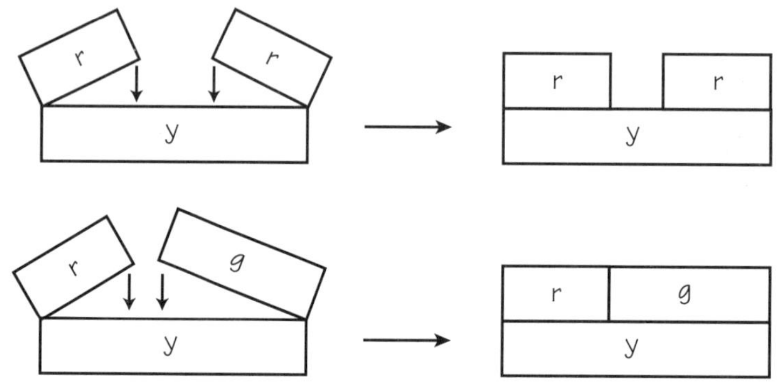

Students may go about their search in various ways. Some may search randomly, while others may have a more organized approach. In Part 1, students should be encouraged to accumulate a good number of examples in each of their lists before trying to formulate any conclusions as to why some combinations form triangles and some don't. As there are numerous examples of both combinations that do and do not work, this should not prove to be too difficult or time consuming.

Some students may choose to investigate why some combinations work while others don't by keeping two of the rods in their triangles constant and experimenting with different choices for the third. The chart at the right shows combinations made with the 3-cm rod (light green) and the 5-cm rod (yellow), and all possible choices for the third rod. Looking at data in this way may help students to see that the third length must be neither too small nor to large. Some students may recognize that the third side must be greater than the difference between the given two sides and less than their sum.

Rod Combination	Triangle?
3,5,1	no
3,5,2	no
3,5,3	yes
3,5,4	yes
3,5,5	yes
3,5,6	yes
3,5,7	yes
3,5,8	no
3,5,9	no
3,5,10	no

There are 220 distinct three-rod combinations that can be chosen from a set of Cuisenaire Rods. Of these 220 combinations, only 125 will form triangles. (In these counts, triangles that are congruent to each other by reflections and/or rotations are not considered to be different triangles.) Ten of the triangles will be equilateral (all sides congruent), 65 triangles will be isosceles but not equilateral (exactly two sides congruent), and the remaining 50 triangles will be scalene (no sides congruent).

The 125 combinations are listed below:

1,1,1	1,2,2	1,3,3	1,4,4	1,5,5	1,6,6
1,7,7	1,8,8	1,9,9	1,10,10		
2,2,2	2,2,3	2,3,3	2,3,4	2,4,4	2,4,5
2,5,5	2,5,6	2,6,6	2,6,7	2,7,7	2,7,8
2,8,8	2,8,9	2,9,9	2,9,10	2,10,10	
3,3,3	3,3,4	3,3,5	3,4,4	3,4,5	3,4,6
3,5,5	3,5,6	3,5,7	3,6,6	3,6,7	3,6,8
3,7,7	3,7,8	3,7,9	3,8,8	3,8,9	3,8,10
3,9,9	3,9,10	3,10,10			
4,4,4	4,4,5	4,4,6	4,4,7	4,5,5	4,5,6
4,5,7	4,5,8	4,6,6	4,6,7	4,6,8	4,6,9
4,7,7	4,7,8	4,7,9	4,7,10	4,8,8	4,8,9
4,8,10	4,9,9	4,9,10	4,10,10		
5,5,5	5,5,6	5,5,7	5,5,8	5,5,9	5,6,6
5,6,7	5,6,8	5,6,9	5,6,10	5,7,7	5,7,8
5,7,9	5,7,10	5,8,8	5,8,9	5,8,10	5,9,9
5,9,10	5,10,10				
6,6,6	6,6,7	6,6,8	6,6,9	6,6,10	6,7,7
6,7,8	6,7,9	6,7,10	6,8,8	6,8,9	6,8,10
6,9,9	6,9,10	6,10,10			
7,7,7	7,7,8	7,7,9	7,7,10	7,8,8	7,8,9
7,8,10	7,9,9	7,9,10	7,10,10		
8,8,8	8,8,9	8,8,10	8,9,9	8,9,10	8,10,10
9,9,9	9,9,10	9,10,10			
10,10,10					

SHELF BRACKETS

- Right triangles
- Area
- Spatial visualization

Getting Ready

What You'll Need

Geoboards, 1 per student

Rubber bands

Geodot paper, page 108

Scissors

Activity Master, page 92

Overview

Students search for all the different-sized right triangles that can be made on the Geoboard. They find the area of each of their triangles, and then use the triangles to solve a problem involving triangular shelf supports. In this activity, students have the opportunity to:

- reinforce understanding of the attributes of right triangles
- use a variety of methods for finding areas of triangles
- use logical reasoning to search for all possible solutions

Other *Super Source* activities that explore these and related concepts are:

Rooftop Triangles, page 22
Hydroponics, page 30
Sal's Similar Sails, page 34

The Activity

On Their Own (Part 1)

Kari is building shelves to hold her stereo, books, and pictures. The shelves are to be various widths, each requiring 2 identical right-triangular wooden support brackets underneath. If Kari has a thick piece of wood measuring 24 in. x 24 in., what are the different-sized right-triangular brackets she can cut from the piece of wood?

- Working with a partner, make as many different-sized right triangles as you can on your Geoboard. Use the Geoboard pegs for the vertices of your triangles.
- Imagine that your Geoboard represents Kari's 24 in. x 24 in. piece of wood and that your triangles are patterns for the shelf brackets. Find the area of each of the shelf brackets you modeled. (Hint: First figure out the dimensions and area represented by each small Geoboard square.)
- Draw each right-triangular bracket on geodot paper and record its area. Compare your drawings to make sure they are all different.
- Be ready to explain how you know you have found all possible solutions.

Thinking and Sharing

Invite pairs of students to take turns posting one of their triangles and telling how they determined its area. Once all the different triangles have been posted and their areas identified, have the class help you find a way to organize the display.

Use prompts like these to promote class discussion:

- How did you go about searching for right triangles?
- How did you decide whether you had found all possible right triangles? How could you convince a classmate that you had found them all?
- How are the triangles you found different from each other?
- How did you find the areas of the triangular brackets you modeled?
- Did anyone use more than one method to find the areas? If so, explain.
- Which areas were easy to find? Which areas were hard to find? Why?

On Their Own (Part 2)

What if... ***Kari wants to construct 4 shelves of varying widths? If each shelf requires a set of 2 identical right-triangular brackets with one length the width of the shelf, how can she make maximum use of the 24 in. x 24 in. piece of wood?***

- Working with your partner, make 2 copies of each right-triangular bracket pattern on geodot paper and cut them out.
- On a clean sheet of geodot paper, connect the outermost set of dots to create a square representing Kari's 24 in. x 24 in. piece of wood.
- Working with your cutouts, find a way to fit 4 pairs of different-sized triangles in the ruled-off geodot square. Start by arranging one pair of identical triangles, then a second pair, a third pair, and finally a fourth pair. The triangles may be reflected and/or rotated as they are placed, but they may not overlap. Try to minimize the amount of wood that would be wasted once the triangular supports are cut out.
- Once you've found an arrangement that you like, trace it onto the paper.
- Calculate the total area used by the 8 brackets in your arrangement, and the area of any leftover wood.
- Be ready to tell about the strategies you used for solving Kari's problem.

Thinking and Sharing

Invite volunteers to present their solutions.

Use prompts like these to promote class discussion:

- What strategies did you use to select the 4 pairs of right-triangular brackets?
- What was the most efficient way of arranging the pairs of triangles to fit in the geodot square? Explain.

- Did you use reflections and/or rotations in placing the right triangles? If so, explain how and why.
- Did anyone form an arrangement that contains rectangles? If so, how are the rectangles related to the triangles?
- How did you calculate the area of any leftover wood in the 24 in. x 24 in. square?

Write a letter to Kari explaining how she could use a Geoboard to find the areas of the triangular supports that can be cut from her piece of wood. Be sure to include instructions for triangles whose sides are not parallel to the edges of the board, as well as for those whose are. Include any diagrams that might be helpful.

Teacher Talk

Where's the Mathematics?

There are 17 different right triangles that can be made using the pegs on the Geoboard as vertices. Of these, the right triangles whose legs are parallel to the sides of the Geoboard are probably the easiest for students to find. To find their areas, students may apply the formula for area of a triangle (Area = (1/2) x base x height), or they may count unit squares, matching pieces of partial squares to form whole and half squares. Students must also realize that each of the small Geoboard squares represents a 6 in. x 6 in. square of the wooden board.

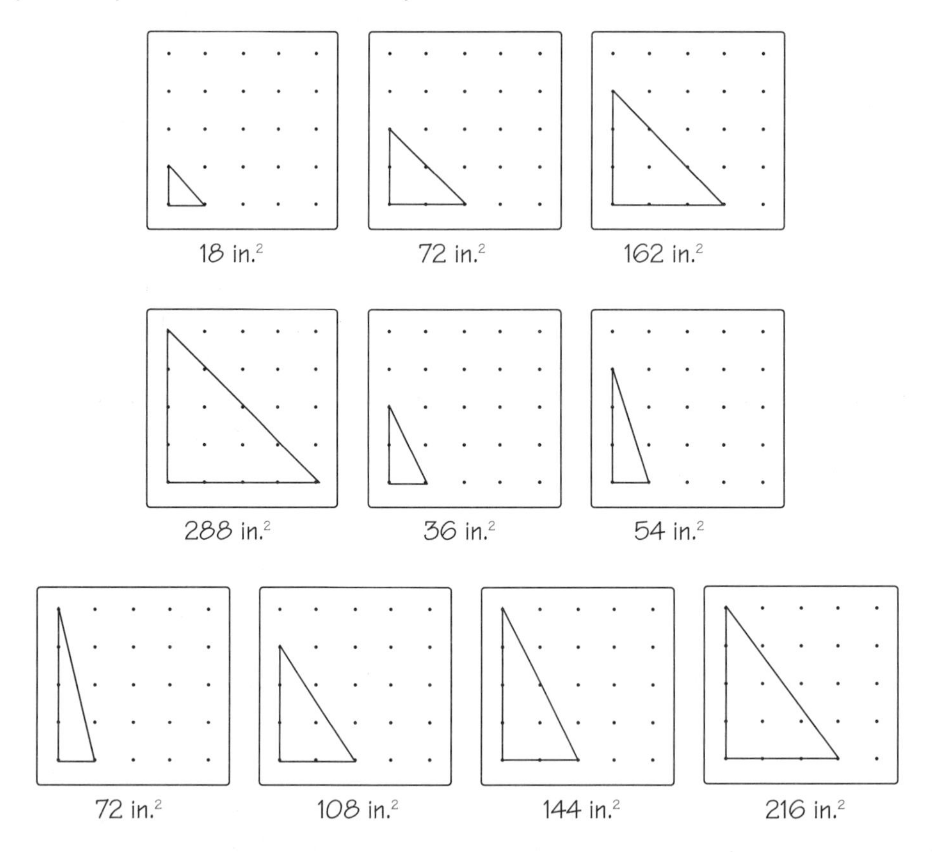

Right triangles whose legs are not parallel to the sides of the Geoboard are shown below with their corresponding areas. Finding these areas may prove to be somewhat challenging for some students.

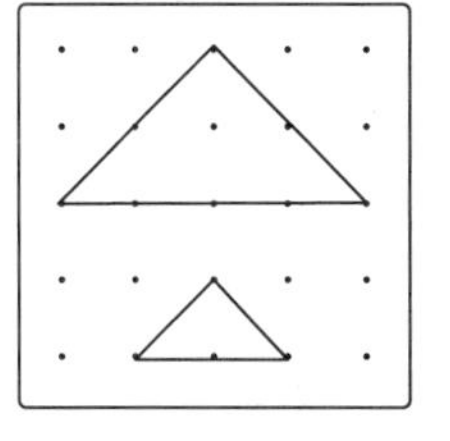

144 in.^{2}, 36 in.^{2}

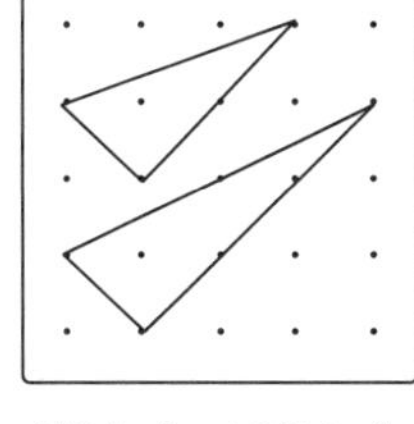

72 in.^{2}, 108 in.^{2}

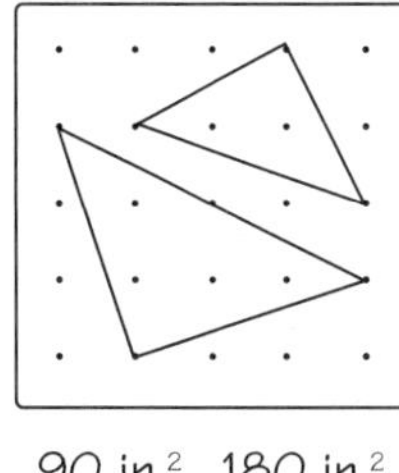

90 in.^{2}, 180 in.^{2}

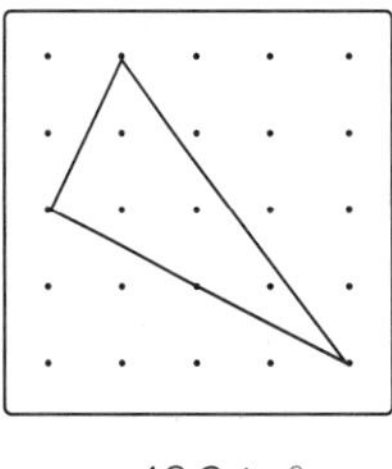

180 in.^{2}

There are several ways students might figure out the area of these triangles. Some students may enclose the triangle in a rectangle that has its sides parallel to the sides of the Geoboard as shown below. The area of the original triangle can then be calculated by finding the area of the rectangle and subtracting the areas of the triangles that surround the original triangle.

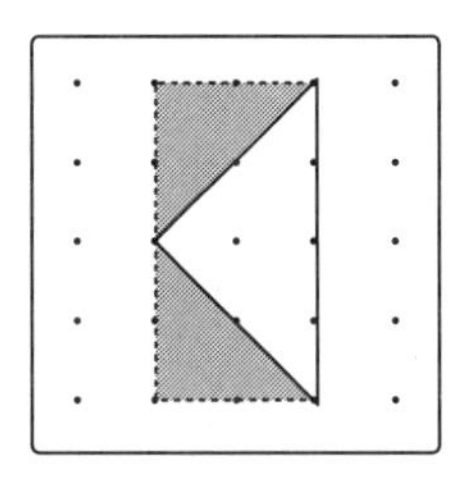

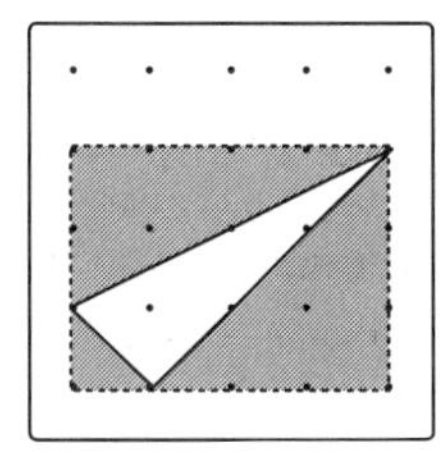

Students may describe other methods for finding areas. Some may be combinations of the techniques described, while others may be totally different techniques that students found useful. Encourage students to describe their strategies, and to demonstrate how they used them to find the areas of their triangles.

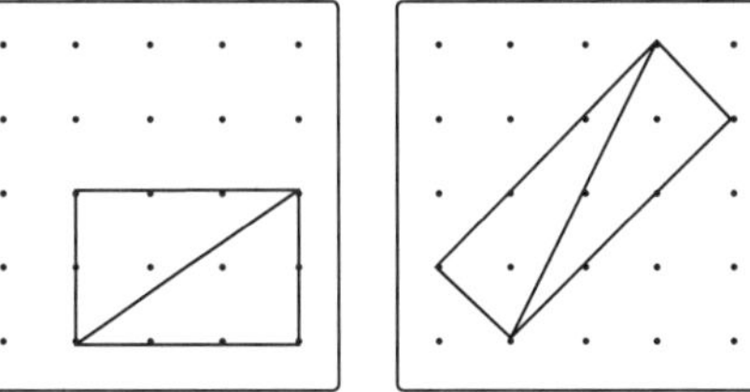

In Part 2, as students work on ways of placing their pairs of right triangles in the Geodot square, they may discover that the most efficient way of arranging each set of identical triangles is in the shape of a rectangle. This helps to avoid wasted space between the triangles.

Once students have chosen the first pair of right triangles to place on the Geodot square, there will be limitations on the dimensions of the three remaining sets of triangles that students may select. Students may come to realize that the rectangles formed from each set of identical triangles are best placed with their sides parallel to the sides of the Geodot square. To find the area of the leftover wood, students can subtract the sum of the areas of the 8 triangular brackets from the area of the original square piece of wood, which is 576 in.^{2}

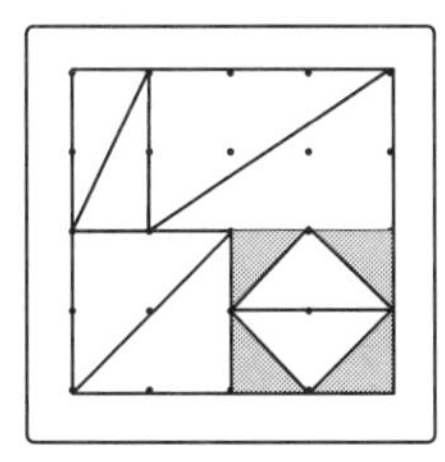

area of leftover wood = 36 in.^{2}

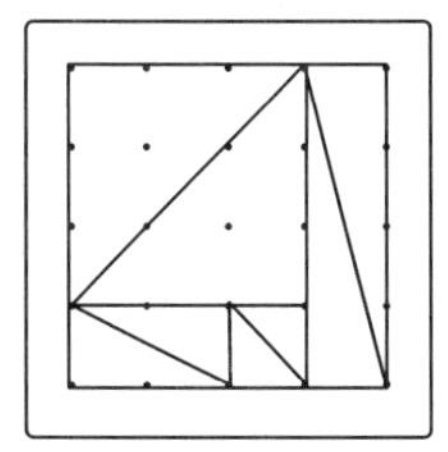

no leftover wood (most efficient use of wooden board)

HYDROPONICS

- Properties of isosceles triangles
- Inscribed and central angles
- Chords and their arcs
- Similarity

Getting Ready

What You'll Need

Circular Geoboards, 1 per student

Rubber bands

Circular geodot paper, page 110

Rulers

Scissors

Activity Master, page 93

Overview

Students search to find all possible isosceles triangles that can be formed on a circular Geoboard. They then look for patterns and relationships among the triangles. In this activity, students have the opportunity to:

- discover that in a triangle, angles opposite congruent sides are congruent
- find the measures of central and inscribed angles in circles
- learn that congruent chords in the same circle intercept congruent arcs

Other *Super Source* activities that explore these and related concepts are:

Rooftop Triangles, page 22

Shelf Brackets, page 26

Sal's Similar Sails, page 34

The Activity

On Their Own (Part 1)

Hydroponics is the science of growing plants in water containing dissolved inorganic nutrients. The plants in their water containers are supported on isosceles triangular frames as they grow. Different-sized plants require different-sized frames. Using the circular Geoboard, how many different triangular frames can you design?

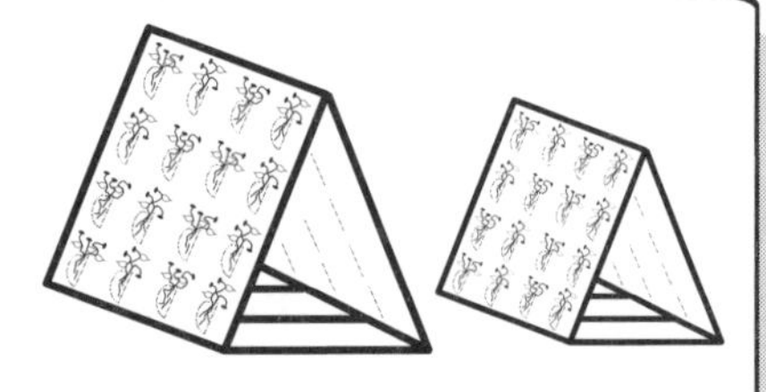

- Working with a partner, create as many different-sized isosceles triangles as you can on your circular Geoboard. Remember that isosceles triangles have at least 2 congruent sides. The vertices of your triangles can be the center peg and 2 pegs on the circle, or they can be 3 pegs on the circle.
- Find the measure of each angle of your isosceles triangles. To do this, first find the measure of the arc formed by any 2 consecutive pegs on the circle. (Remember that the measure of the entire circle is 360°.) Then use the following formulas to help:
    - ◆ The measure of a central angle is equal to the measure of the arc it intercepts.

- The measure of an inscribed angle is half the measure of the arc it intercepts.

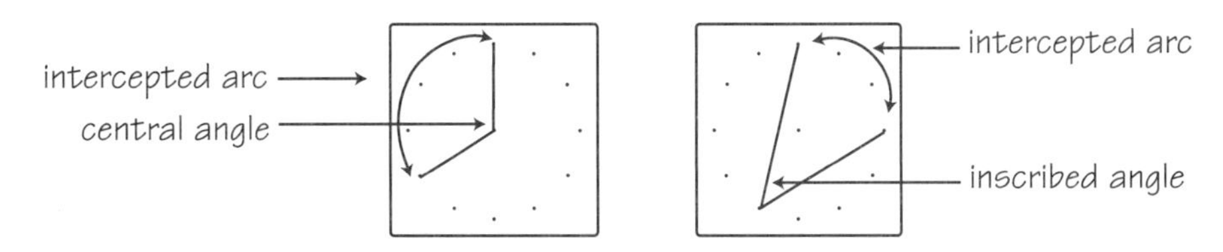

- Draw each isosceles triangle on circular geodot paper and record its angle measures.
- What observations can you make about your triangles?

Thinking and Sharing

Invite volunteers, one at a time, to recreate one of their triangles on their Geoboard and display it on the chalk rail. Have them label the measure of each angle of their triangle. Continue this process until students are confident that all isosceles triangles have been displayed.

Use prompts like these to promote class discussion:

- How many isosceles triangles did you find?
- Do you think you've found all of the isosceles triangles that can be made on the circular Geoboard? How do you know?
- How did you decide whether the side lengths were congruent?
- What observations did you make about your triangles?
- How would you describe the properties of an isosceles triangle?
- Is an equilateral triangle also isosceles? Explain.
- Is an isosceles triangle also equilateral? Explain.
- If two sides of a triangle are congruent, what can you say about the angles of the triangle? Explain.
- Did you notice any other patterns? If so, tell about them.

On Their Own (Part 2)

What if... ***the hydroponics frames were to be sold in sets containing one small triangular frame and one large triangular frame? How would you select the pairs of frames that would be packaged together?***

- Working with your partner, cut out each of your isosceles triangles.
- Think of a way to group your "triangular frames" into sets of two for packaging.
- Investigate the properties shared by the triangles in each of your sets. Compare angle measures, types of angles, and side lengths.

Thinking and Sharing

Invite several pairs of students to share their solutions and explain why they paired the triangles as they did.

Use prompts like these to promote class discussion:

- What strategies did you use to pair the triangular frames?
- What relationships exist between the triangles in each pair?
- What did you discover about the angle measurements of the triangles in each set?
- What did you discover about the side lengths of the triangles in each set?
- How would you classify each set of triangular frames?

List at least five geometric properties that you learned about in this activity that you didn't know before. Use diagrams to help illustrate each property.

Teacher Talk

Where's the Mathematics?

In creating isosceles triangles on the circular Geoboard, students can investigate the angles that are opposite the congruent sides and discover that they also are equal in measure. Using the circular Geoboard in developing this concept also provides students an opportunity to learn about central angles, inscribed angles, and chords of circles. There are ten possible isosceles triangular frames that can be modeled on the Geoboard. Students may need to rotate their drawings to compare their triangles to the ones shown.

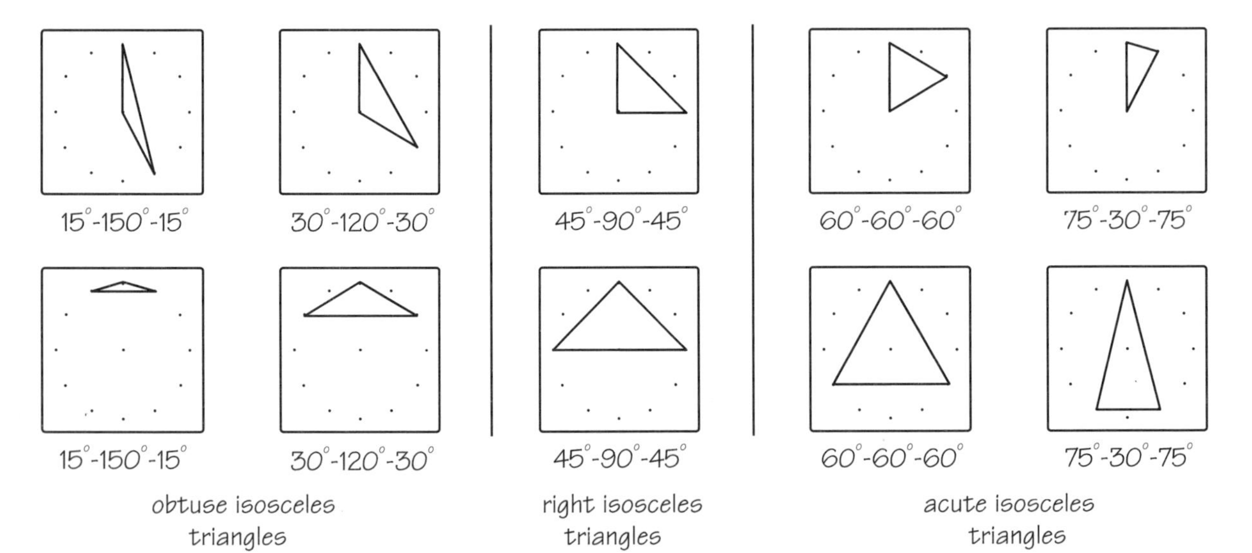

Students may have different ways of describing how they know that sides of their triangles are congruent. For the five isosceles triangles whose vertices are points on the circle, students may notice that the sides they think are congruent intercept congruent arcs, or arcs containing the same number of pegs. Point out that these sides are chords of the circle, and what they're discovering is that congruent chords of the same circle intercept congruent arcs.

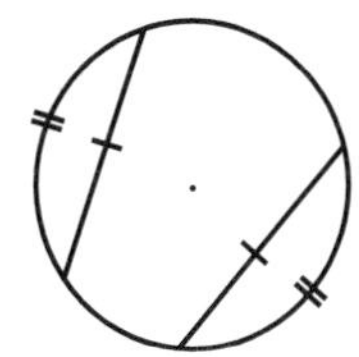

congruent chords and congruent arcs

The five other isosceles triangles are formed using the center of the circle as one of the vertices. These triangles each have two sides that are radii of the circle and are therefore congruent. You may want to remind students that a circle is defined as the set of points in a plane that are a given distance (the radius) from a given point (the center).

Before working on this activity, students may not have realized that equilateral triangles are also isosceles. Help them to see that if three sides of a triangle are congruent, certainly any two of its sides are congruent. It is also possible for a triangle labeled "isosceles" to be equilateral. By saying that a triangle has two congruent sides, the possibility of its having three congruent sides is not ruled out.

Students should notice that the two congruent angles of the triangle are always opposite the two congruent sides of the triangle.

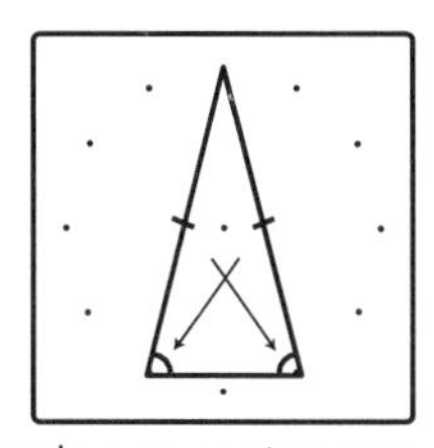

congruent angles opposite congruent sides

They may also recognize that since the measure of the arc between any two consecutive pegs is 30°, inscribed angles on their Geoboard will be multiples of 15° and central angles will be multiples of 30°.

In determining which pairs of isosceles triangular frames might be packaged together, students may select those triangles that have congruent angle measurements. These pairs of triangles are similar; therefore, not only are their corresponding angles congruent, but their corresponding side lengths are also proportional, enabling them to fit evenly, one inside the other. The pair whose angles measure 30°-120°-30° are not only similar, but are also congruent.

SAL'S SIMILAR SAILS

- Areas and perimeters of triangles
- Non-standard measurement
- Properties of similar triangles
- Spatial visualization

Getting Ready

What You'll Need

Tangrams, 1 set per pair

Scissors

Activity Master, page 94

Overview

Students build as many different-sized triangles as they can using one set of Tangram pieces, and compare their side lengths, perimeters, and areas. In this activity, students have the opportunity to:

- create and compare different-sized triangles
- use spatial reasoning
- reinforce the concept of similarity
- determine and compare angle measures, side lengths, perimeters, and areas of similar triangles

Other *Super Source* activities that explore these and related concepts are:

Rooftop Triangles, page 22

Shelf Brackets, page 26

Hydroponics, page 30

The Activity

On Their Own (Part 1)

Sal is a sail maker at a marina in a small coastal village. He constructs sails using combinations of different-sized pieces of canvas. If Sal has 7 pieces of canvas shaped like Tangram pieces, how many different-sized sails can he make?

- Working with a partner, make a model of a triangular sail using 2 or more Tangram pieces.
- Find the number of units in the area of your model. Let the area of the smallest Tangram triangle represent 1 unit.
- Trace your model on a sheet of paper. Include the outlines of the shapes you used to build it. Record its area.
- Continue to make and record triangles until you have made models of all the different-sized triangular sails that you can. Then cut out each triangle.
- Compare the triangles and be ready to discuss their similarities and their differences.

Thinking and Sharing

Invite a pair of students to display one of their models and record its area on the chalkboard. Ask if other pairs built models that have the same area but are made from different Tangram pieces or different arrangements of pieces. Post these triangles with the first. Continue this process for models with different areas until students agree that all solutions have been displayed.

Use prompts like these to promote class discussion:

- What do you notice about the models?
- How do models with the same area differ? How are they alike?
- How did you go about searching for all the different-sized triangles that could be made?
- Are you convinced that it's not possible to make triangles with areas other than those displayed? Why or why not?
- Which Tangram pieces did you use most often? Which pieces were the most difficult to use?
- How are the triangular sails alike? How are they different?

On Their Own (Part 2)

What if... *Sal had to stitch roping onto each side of the sails to help them maintain their shape? How much rope would Sal need for each sail?*

- Using the length of the shorter side of the smallest Tangram triangle to represent 1 unit of length, find the side lengths of each triangle from Part 1.
- Find the perimeter of each triangle.
- For each model, record the lengths of the 3 sides and the perimeter.
- Pick any 2 of your triangles and compare them. Compare side lengths, perimeters, and areas. What relationships can you find? Test these relationships using a different pair of triangles. What generalizations can you make about the triangular sails?

Thinking and Sharing

Have students record their measurements on the posted triangles. If there is disagreement, have students work together to remeasure and, if necessary, refigure any of the lengths.

Use prompts like these to promote class discussion:

- How did you go about finding the side lengths of your triangles?
- Which side lengths were easy to find? Why?
- Which side lengths were hard to find? Why?
- How did you find the perimeters of your triangles?
- What relationships did you find?

- Did you notice any relationships among the ratios of corresponding sides, perimeters, and areas? If so, tell about them.
- What generalizations can you make about the triangular sails?

Write a brief letter to Sal, listing and explaining all the different properties you discovered about the possible sails he can construct. Include any diagrams that might be helpful.

Teacher Talk

Where's the Mathematics?

There are 5 different-sized triangles that can be made using two or more pieces from a set of Tangrams. The examples shown below are not drawn in proportion and, therefore, the labels S, M, and L are used to indicate the small, medium, and large triangle pieces of the Tangram set.

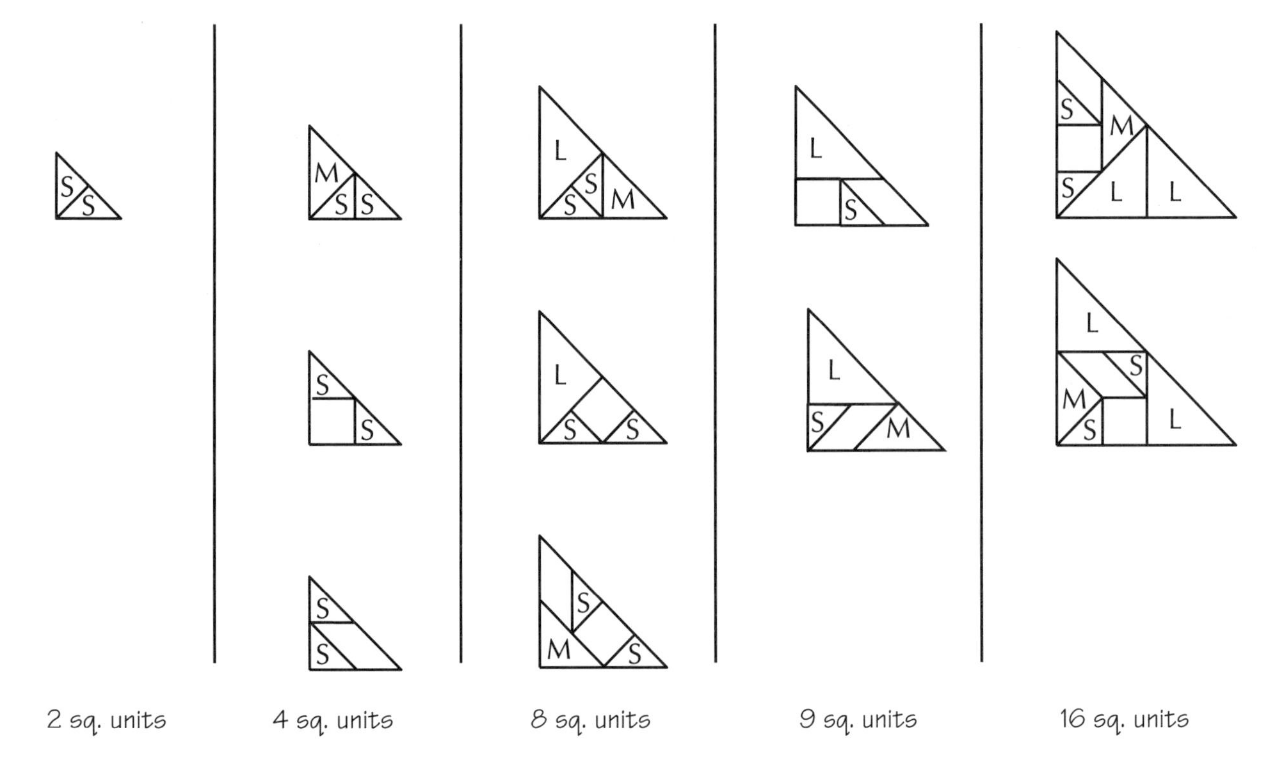

2 sq. units | 4 sq. units | 8 sq. units | 9 sq. units | 16 sq. units

With the exception of the smallest triangle, there is more than one way to make each of the other triangles.

Students should find that each of the triangular sails will be an isosceles right triangle. The triangles contain two congruent sides that are perpendicular to each other and thus form a right angle. Since all of the triangular sails are isosceles right triangles, they are similar to each other; that is, the triangles will have the same shape but will not necessarily be the same size. Their corresponding angles are congruent and their corresponding sides are in proportion. Encourage students to express these relationships mathematically.

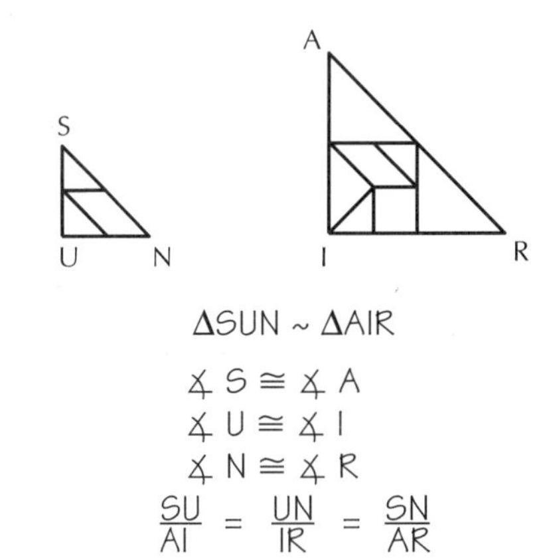

Working with the cutout triangles, students can place one triangle on top of another to see that corresponding angles are congruent. Corresponding side lengths can be matched and compared in a similar way. In the triangles above, $\overline{AI}$ is twice as long as its corresponding side, $\overline{SU}$, $\overline{IR}$ is twice as long as its corresponding side, $\overline{UN}$, and $\overline{AR}$ is twice as long as its corresponding side, $\overline{SN}$. The ratio of the side lengths of ΔSUN to that of ΔAIR is 1:2. It is also true that the ratio of their perimeters is 1:2.

Finding side lengths in terms of the length of the side of the small Tangram triangle piece is easy for those triangular sails with side lengths that are multiples of the unit length. For those side lengths that are not multiples, students may use the Pythagorean Theorem: $a^2 + b^2 = c^2$, where a and b are the lengths of the legs of the right triangle, and c is the length of the hypotenuse. For students who have not learned about irrational numbers, using calculators to obtain decimal approximations for these lengths may be appropriate.

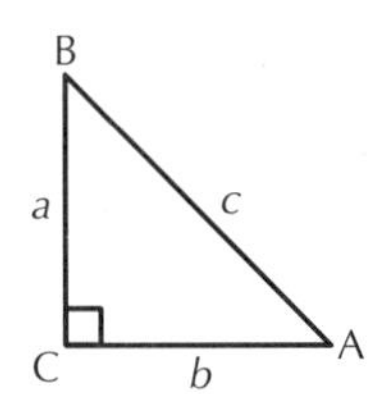

As students investigate the relationships that exist among their triangles, they may notice that the ratio between the perimeters of two triangles is equal to the ratio of their corresponding sides. They may think the same will be true of the ratio of the areas, but may surprised to find that the area ratio is actually the square of the ratio of corresponding sides. For example, in the diagram above, the ratio of the area of ΔSUN to ΔAIR is 1:4, whereas the ratios of their perimeters and of their corresponding side lengths are 1:2. Students can generalize that if two triangles are similar, and the ratio of corresponding side lengths is a:b, then the ratio of their perimeters is a:b, and the ratio of their areas is a^2:b^2.

Investigating Polygons

1. *Geoboard Challenge, page 39* (*Geoboards*)
2. Braille Puzzles, page 43 (Tangrams)
3. Circuit Boards, page 47 (Tangrams)
4. Star Search, page 51 (Circular Geoboards)

The first two activities in this cluster are ideal for reinforcing basic vocabulary about polygons; in the first activity students need to recognize terms given to them, and in the second activity students need to come up with those terms themselves. The last two activities would work either as introduction to or reinforcement of the concepts of, respectively, convex and concave polygons, and diagonals of polygons.

1. Geoboard Challenge (*Reinforcing geometric classification*)

This activity helps students reinforce their understanding of vocabulary relating to polygons. *On Their Own* Part 1 assumes a familiarity with these terms: Isosceles triangle, obtuse angle, right angle, congruent angles, rectangle, square, hexagon, and parallel. *Where's the Mathematics?* discusses (on page 42) the distinction between convex and concave polygons.

Where's the Mathematics? (page 41) suggests a connection to the theorem stating that angles opposite the congruent sides of an isosceles triangle are congruent. It also suggests (on page 42) a connection to the Pythagorean Theorem.

2. Braille Puzzles (*Reinforcing geometric vocabulary*)

In this activity, students come up with the vocabulary needed to describe geometric figures. This contrasts with *Geoboard Challenge*, in which students simply need to recognize vocabulary given to them.

Teachers may want to introduce this activity with a quick review of vocabulary that the class has studied. Some of the terms likely to be helpful are listed in *Where's the Mathematics?*, page 45.

3. Circuit Boards (*Exploring convex and concave polygons*)

In this activity students learn the distinction between concave and convex polygons. In building their polygons, students have an opportunity to reinforce geometric vocabulary while describing their polygons.

On Their Own Part 1 outlines the straight-segment method for determining whether a polygon is concave or convex; therefore, the activity would work well as an introduction to the topic.

Where's the Mathematics? (page 49) gives several solutions and lists some vocabulary students might use in describing their polygons.

4. Star Search (*Exploring diagonals of a polygon*)

In this activity, students build diagonals in various polygons and discover relationships among the numbers of vertices, sides, diagonals, and triangles formed by diagonals from a single vertex.

On Their Own Part 1 assumes that students can make a diagonal in a given polygon. If students have not worked with diagonals, the teacher may want to precede the activity with a quick preactivity on drawing diagonals.

GEOBOARD CHALLENGE

- Polygons
- Properties of geometric shapes
- Spatial reasoning

Getting Ready

What You'll Need

Geoboards, 1 per student

Rubber bands

Geodot paper, page 109

Scissors

Activity Master, page 95

Overview

Students play a game in which they try to make Geoboard polygons that fit given descriptions. In this activity, students have the opportunity to:

- explore attributes of different polygons
- use mathematical reasoning to determine whether or not it is possible to make a shape having a particular set of attributes
- use mathematical language to describe polygons and their attributes

Other *Super Source* activities that explore these and related concepts are:

Braille Puzzles, page 43

Circuit Boards, page 47

Star Search, page 51

The Activity

On Their Own (Part 1)

Audrey has invented a Geoboard game called Geoboard Challenge. She has written 14 clues, each describing a shape that may or may not be possible to create on a Geoboard. To play the game, you must try to create the shape on the Geoboard, if it exists. Play Geoboard Challenge and see how many points you can score!

- For each description below, decide if it is possible or impossible to create the shape on your Geoboard. If you find a solution, award yourself 1 point. If you can find more shapes that fit the description, award yourself another point for each solution.
- Record your solutions on geodot paper. Label them with the number of the corresponding clue and cut them out. If you think it is impossible to make the shape described, be ready to explain why.

Geoboard Challenge

1. Find a triangle that contains 5 interior pegs.
2. Find a triangle that contains 6 interior pegs.
3. Find a triangle that contains 7 interior pegs.
4. Find an isosceles triangle that has one obtuse angle.

5. Find an isosceles triangle that has a right angle.
6. Find an isosceles triangle that has no congruent angles.
7. Find a rectangle that contains 3 interior pegs.
8. Find a rectangle that contains 5 interior pegs.
9. Find a square that has side lengths greater than 2 units but less than 3 units.
10. Find a square that has an area of 6 square units.
11. Find a hexagon that has no parallel sides.
12. Find a hexagon that has 3 parallel sides.
13. Find a hexagon that has no congruent sides.
14. Find a hexagon that has all sides congruent.

Thinking and Sharing

On the chalkboard, create 14 columns with headings corresponding to each description. Ask students to post their geodot drawings in the appropriate columns. Have them eliminate duplicate solutions. For those descriptions that students determined to be impossible to satisfy, have students explain why they reached that conclusion. If students disagree, allow them time to work together to prove or disprove their conclusions.

Use prompts like these to promote class discussion:

- What discoveries did you make while searching for solutions?
- Which descriptions for shapes were the easiest to satisfy using the Geoboard? Which were the most difficult? Why?
- How many points did you score in the game?
- How did you decide when it was impossible to find a solution?
- Do you think that some of the shapes that you decided were impossible to make on the Geoboard would be possible to draw on plain paper?

On Their Own (Part 2)

What if... ***you wanted to make up your own version of Geoboard Challenge to play with a friend? What clues would you create?***

- Create a list of 10 clues for shapes that may or may not be possible to make on the Geoboard. Make up an answer key. Keep it hidden from your partner.
- Exchange clues with your partner.
- Take turns trying to solve a clue from each other's list. Score 1 point for each solution you can find for a given clue. Score 1 point if you correctly identify a description as impossible to satisfy and are able to explain why. Score 1 point each time you stump your partner.
- Keep track of your score and your partner's score to see who scores the most points.

Thinking and Sharing

Have students take turns sharing some of the clues they wrote and describing what happened during their games.

Use prompts like these to promote class discussion:

- How did you go about making up your clues?
- What was hard about creating the clues?
- Was it harder to create descriptions for shapes that were possible to make or for shapes that were not possible? Explain.
- Did you and your partner have any duplicate clues or clues that had the same solutions? If so, tell about them.
- Did you or your partner find a solution that was different from the one or ones on the answer key? How were the solutions alike? How were they different?
- Were there descriptions that you thought could not be satisfied that could? If so tell about your thinking.
- Who won the game? Looking back, why do you think that person was able to score more points?

Write a clue that has exactly 5 solutions. Draw your solutions on geodot paper. Then choose one of the solutions and write a clue for which the chosen solution is the only solution.

Teacher Talk

Where's the Mathematics?

The first 6 descriptions focus on triangles and some of their Geoboard properties. Possible solutions for clues 1 and 2 are shown below. As they experiment with different-sized triangles, students may realize that the maximum number of pegs that can be enclosed in a triangle is 6.

a triangle with 5 interior pegs

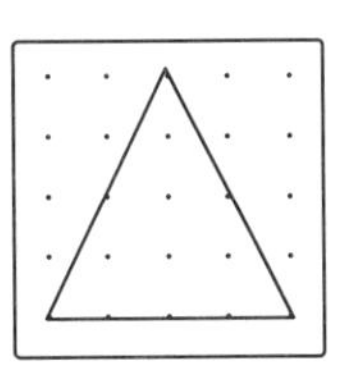

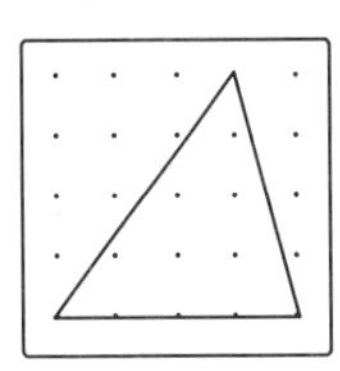

a triangle with 6 interior pegs

Students should be able to make several different isosceles triangles that contain either an obtuse angle or a right angle. Possible solutions are shown below. In studying their solutions to these clues, students may notice that if two sides of a triangle are congruent, the angles opposite these sides are also congruent. This fact makes it impossible to build on the Geoboard – or even sketch on paper – an isosceles triangle that has no congruent angles.

an isosceles triangle with 1 obtuse angle

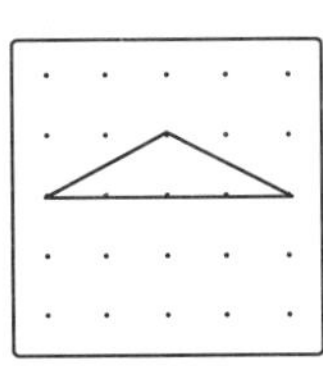

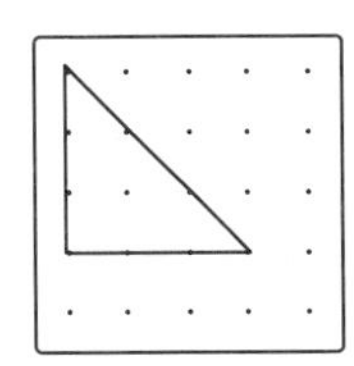

an isosceles triangle with 1 right angle

Two rectangles that contain 3 interior pegs are shown below. Students should be encouraged to compare the lengths of the sides of their rectangles to verify that they are not congruent. The rectangle that contains 5 interior pegs is actually a square, also shown below.

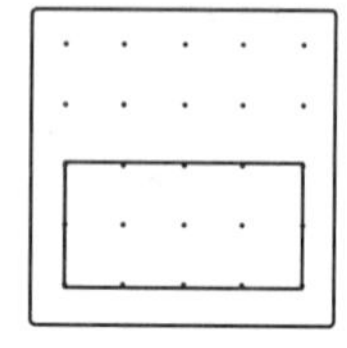
a rectangle with 3 interior pegs

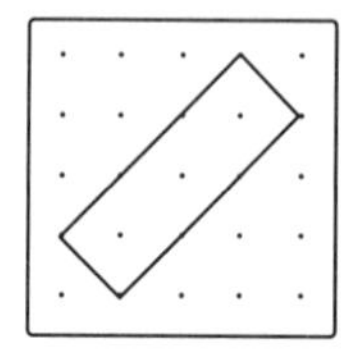
a rectangle with 3 interior pegs

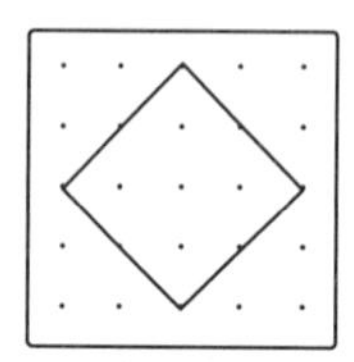
a rectangle with 5 interior pegs

To find a square with side lengths greater than 2 units but less than 3 units, students must realize that the diagonal distance between two pegs is greater than the horizontal or vertical distance between two pegs. The Pythagorean Theorem can be used on the right triangle marked within the square below to verify that the lengths of the sides of this square are between 2 and 3 units.

sides longer than 2 units but shorter than 3 units

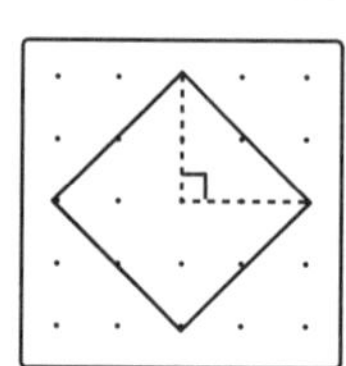

$$2^2 + 2^2 = (\text{side})^2$$
$$4 + 4 = (\text{side})^2$$
$$8 = (\text{side})^2$$
$$\sqrt{8} = \text{side}$$

$$\sqrt{4} < \sqrt{8} < \sqrt{9}$$
$$2 < \sqrt{8} < 3$$

In their attempt to build a square that has an area of 6 square units, students may start with a 2-by-3 rectangle, and try to adjust the sides of this rectangle to form a square. Students should discover that it is impossible to make this square on the Geoboard.

For the clues about hexagons, students will need to consider both convex hexagons (those containing angles whose measures are all less than 180°) and concave hexagons (those containing some angles whose measures are greater than 180°). Parallel sides may be easily recognizable if they are located on vertical or horizontal lines of pegs. Other parallel sides may be harder to visualize. Students may want to extend the sides to verify that they remain the same distance apart, or check to see that the segments rise or fall at the same rate. For example, in the third diagram below, each of the three parallel sides has a vertical rise of 1 unit for every horizontal run of 2 units.

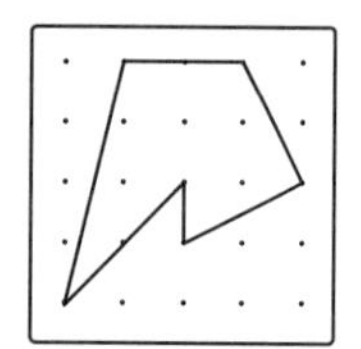
a hexagon with no parallel sides

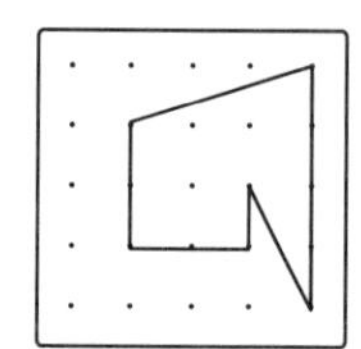
a hexagon with 3 parallel sides

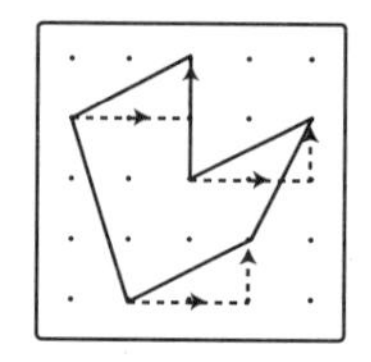
a hexagon with 3 parallel sides

Although it is possible to build many hexagons that have no pairs of congruent sides, students may find that building one that has all sides congruent is a bit more challenging.

Students' explanations as to why certain descriptions cannot be satisfied may reflect the depth of their understanding of geometric concepts and the limitations of the Geoboard. As they create their own clues and solve each other's in the second part of the activity, they have a chance to reinforce this understanding and to use mathematical reasoning to justify and verify their thinking.

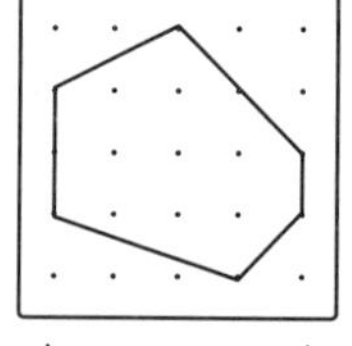
a hexagon with no congruent sides

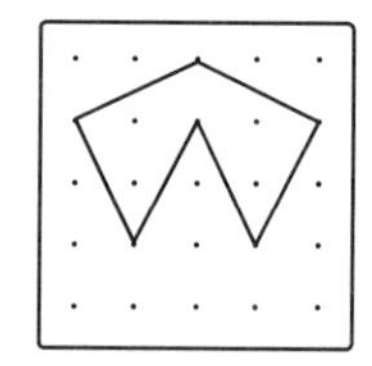
a hexagon with all sides congruent

BRAILLE PUZZLES

- Polygons
- Mathematical vocabulary
- Writing and following instructions
- Spatial visualization

Getting Ready

What You'll Need

Tangrams, 1 set per student

Markers

Protractors

Metric rulers

Activity Master, page 96

Overview

Students create Tangram designs, describe them in writing, and try to recreate each other's designs from the written instructions. They then describe the cuts and rotations a jigsaw would need to make to cut around the perimeter of their designs. In this activity, students have the opportunity to:

- analyze properties of polygons
- write mathematical descriptions
- write and follow instruction
- measure distances and degrees of rotation

Other *Super Source* activities that explore these and related concepts are:

Geoboard Challenge, page 39

Circuit Boards, page 47

Star Search, page 51

The Activity

On Their Own (Part 1)

Michel works for a company that manufactures jigsaw puzzles for the visually impaired. His job is to write step-by-step instructions for assembling the puzzles. The instructions are then translated into Braille for the visually impaired. How accurately would you be able to write these kinds of instructions?

- Sit back-to-back with your partner (or with a barrier between the two of you) so that you will not be able to see each other's work. Make a puzzle design using at least 5 of the 7 pieces from a set of Tangrams.
- On a piece of paper, record your puzzle design, including the placement of the individual Tangram pieces used. This recording will become the answer key to your puzzle.
- On a second sheet of paper, write the step-by-step instructions for making your puzzle design. Use mathematical terms and be concise. Do not use any drawings.
- Exchange puzzle instructions with your partner. Try to follow your partner's instructions for making his or her puzzle design.
- Once you and your partner have finished the puzzles, compare the puzzles to their answer keys. Discuss how the designs compare and how the written instructions might be improved.

Thinking and Sharing

Have students share their puzzles and discuss what was easy and/or what was difficult about writing and following their instructions.

Use prompts like these to promote class discussion:

- What was hard about writing the step-by-step instructions? What was easy?
- Are some Tangram pieces more difficult to describe than others? Which ones and why?
- How did you use the properties of the Tangram pieces to help your partner know how to place each piece correctly?
- Did your partner's instructions name all the Tangram pieces that were needed for the puzzle? If not, how did you know what pieces to use?
- Were there any parts of your partner's instructions that were hard to understand? If so, tell about them.
- When you followed your partner's instructions for assembling his puzzle, did your puzzle design match your partner's answer key exactly? If not, why not?

You may want to make a list on the chalkboard of mathematical vocabulary words used by students in their instructions and during the discussion. Encourage students to provide informal definitions or diagrams illustrating the terms they used.

On Their Own (Part 2)

What if... ***Michel is asked to write instructions for the jigsaw machine that is used to cut the frame of the finished puzzle design? His instructions will then be programmed into the machine, telling the jigsaw how to move around the border of the puzzle to make the appropriate cuts. How might he do this?***

- Sitting back-to-back with your partner, make a new puzzle design using at least 5 of the 7 pieces from a set of Tangrams.
- On a piece of paper, trace the outline, or "frame," of your puzzle design. This recording will become the answer key to your programming instructions.
- On a separate sheet of paper, write the commands you would use to program the jigsaw machine to cut the frame for your puzzle. To do this, select one vertex of your puzzle design as the starting point. Then, using a ruler and protractor as measuring tools, write a series of step-by-step commands that generate a path around the perimeter of the puzzle design, arriving back at the starting point.

starting point

- Exchange instructions with your partner. Try to follow your partner's instructions for making his or her puzzle frame.
- Once you and your partner have finished the frames, compare them to the answer keys. Discuss how the frames compare and how the written instructions might be improved.

Thinking and Sharing

Have students share their puzzle frames and instructions and tell about what happened when they tried to follow each other's instructions.

Use prompts like these to promote class discussion:

- How did you go about writing your commands?
- What was hard about writing the commands? What was easy?
- What mathematical vocabulary words did you use in the commands? Explain, define, or illustrate the terms.
- How did you use the rulers and protractors to figure out the commands?
- Were there any parts of your partner's instructions that were hard to follow? If so, tell about them.
- When you followed your partner's commands for cutting around the design, did your frame design match your partner's answer key exactly? If not, why not?

Make a list of 10 mathematical terms that you think are helpful in describing Tangram designs. Define each of the terms, including diagrams where helpful.

Teacher Talk

Where's the Mathematics?

In attempting to write concise, accurate mathematical descriptions of a puzzle design, students learn to recognize the importance of communicating mathematical ideas using vocabulary that is universally understood. They also strengthen their skills in translating spatial relationships into mathematical statements. The direct feedback provided by their partner's efforts to follow their instructions helps to reinforce the importance of mathematical language and clarity.

Some students may benefit from a review of some common mathematical terminology that they might use in writing their descriptions. Terms like isosceles, equilateral, vertex, parallel, perpendicular, trapezoid, parallelogram, adjacent sides, and hypotenuse may be useful in describing Tangram designs.

When designing their Tangram puzzles, students may choose to create shapes of recognizable objects or they may decide to make an abstract geometric design. One type of puzzle is not necessarily easier to write about than the other. Students may discover that not everyone has the same mental picture of common objects.

Students may find that writing clear mathematical statements can be quite challenging. If students try to recreate their own designs from their step-by-step directions, they may be able to clarify or simplify their instructions or add missing steps. Some students may find that they need to simplify or modify their puzzle design in order to write the step-by-step instructions more easily.

Here is an example of a puzzle design a student might make, along with instructions for its assembly.

My puzzle looks like a house with a chimney. The main part of the house is made of two large triangles joined along their hypotenuses. The legs of these right triangles run horizontally and vertically. The roof of the house is made up of two small triangles joined leg to leg to form an isosceles right triangle. Its hypotenuse sits on top of the square base of the house from vertex to vertex. The chimney is a parallelogram. One of its short sides is attached to the hypotenuse of the small triangle on the right, and one of its longer sides is a straight line extension of the vertical side of the house.

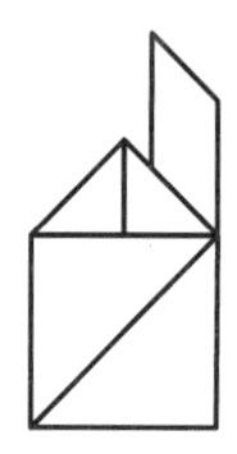

To write instructions for programming the jigsaw cutting machine to cut the frame of their puzzle, students may want to think about their frame as a "pathway" that they might take to "walk around" their design. They will then need to think about directions like "turn right" or "turn left" and use a protractor to measure the angles of rotation. These directions can be replaced with more mathematical terms such as "rotate clockwise" or "rotate counter-clockwise" if students feel comfortable with the vocabulary.

Here is an example of commands for programming the jigsaw cutting machine to cut out the house puzzle above:

Start at the lower left-hand corner of the house. Make a straight line cut 7.1 cm long. Rotate clockwise 45° and make a straight line cut 5 cm long. Rotate clockwise 90° and make a straight line cut 1.5 cm long. Rotate counter-clockwise 135° and make a straight line cut 5 cm long. Rotate clockwise 135° and make a straight line cut 3.5 cm long. Rotate clockwise 45° and make a straight line cut 12.1 cm long. Rotate clockwise 90° and make a straight line cut 7.1 long, ending up at your starting point.

CIRCUIT BOARDS

- Comparing polygons
- Concavity, convexity
- Spatial reasoning

Getting Ready

What You'll Need

Tangrams, 2 sets per student

Uncooked spaghetti

Scissors

Activity Master, page 97

Overview

Students try to create as many different convex polygons as possible using first one and then two sets of Tangrams. In this activity, students have the opportunity to:

- strengthen spatial reasoning
- develop strategies to determine whether a polygon is concave or convex
- compare polygons

Other *Super Source* activities that explore these and related concepts are:

Geoboard Challenge, page 39
Braille Puzzles, page 43
Star Search, page 51

The Activity

On Their Own (Part 1)

An electronics company manufactures circuit boards in convex shapes. The board must be convex so that any two points on it can be joined with a straight segment of wire lying entirely on the surface of the board. What circuit board shapes can they design using a set of Tangrams?

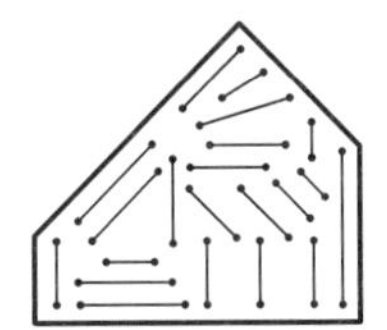

- Work with a partner. Using the complete set of 7 Tangram pieces, design as many different convex polygon shapes as you can to represent the circuit boards.
- Use a piece of uncooked spaghetti to check that your shapes are convex. Here's how: If you choose any two points on the shape and connect them using a piece of spaghetti, and the portion of the piece of spaghetti that lies between the two points remains inside the shape or lies along one of its borders, the polygon is convex. If any portion of the piece of spaghetti between the two points falls outside the shape, the polygon is concave.
- Trace around the border of each convex shape to record your circuit boards. Then cut them out and sort them according to the number of sides they have.
- Compare the cutout circuit board shapes to make sure they are all different. Be ready to discuss the shapes that you found.

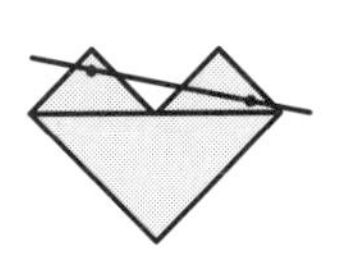

Thinking and Sharing

Ask students what the least and greatest number of sides were that they found. Write this range of values on the chalkboard and invite students to post their recordings under the appropriate column headings: *3-sided polygons, 4-sided polygons, 5-sided polygons, 6-sided polygons.*

Use prompts like these to promote class discussion:

- What strategies did you use to design the convex shapes for the circuit boards?
- What did you notice about the shapes that could be made?
- What was difficult about the activity? What was easy? Why?
- What methods did you use to check that the polygons were all different?
- Do you think there are solutions that have more than 6 sides? Why or why not?
- Do you think you found all possible convex polygons? Why or why not?
- Do you think there are more, fewer, or the same number of concave polygons as convex polygons that can be made using a 7-piece Tangram set? Explain.

On Their Own (Part 2)

What if... *the electronics company wants to expand its production capabilities? If it now can design circuit boards based on shapes formed from 2 sets of Tangrams, what larger convex circuit boards can they produce?*

- Using 2 sets of Tangram pieces, design as many different convex polygon shapes as you can to represent the new circuit boards.
- Check using a piece of spaghetti to make sure your shapes are convex.
- Trace around the border of each convex shape to record your circuit boards. Then cut them out and sort them. Compare the cutout circuit board shapes to make sure they are all different.
- Compare your new shapes to the shapes you made in Part 1. What observations can you make?

Thinking and Sharing

Invite students to add their findings to the existing chart on the chalkboard.

Use prompts like these to promote class discussion:

- What strategies did you use to design the shapes for the circuit boards?
- Did anyone use the shapes from the first activity to help find new shapes? If so, explain what you did.
- Do you think you found all possible convex polygons? Why or why not?
- What methods did you use to check that the polygons were all different?
- How did the new circuit board shapes compare to those from Part 1?

Describe the difference between a convex polygon and a concave polygon. Then state two methods that could be used to determine whether a polygon is convex or concave.

Teacher Talk

Where's the Mathematics?

As students work on designing circuit boards in convex shapes, they have an opportunity to explore the geometric relationships that exist among the Tangram pieces. They may be surprised to find that there is a limited number of ways to arrange the pieces into a convex polygon.

Students will go about their search for convex shapes in a variety of ways. Some may search randomly, while others may use a more systematic approach. One strategy might involve moving pieces in existing solutions to form new solutions. For example, the square below can be split on its diagonal and rearranged to form the large triangle. Then, by shifting one of the larger triangular pieces to a new location, a trapezoid shape is formed. Then, by reflecting that same triangle, a parallelogram shape is formed. All of these solutions are convex.

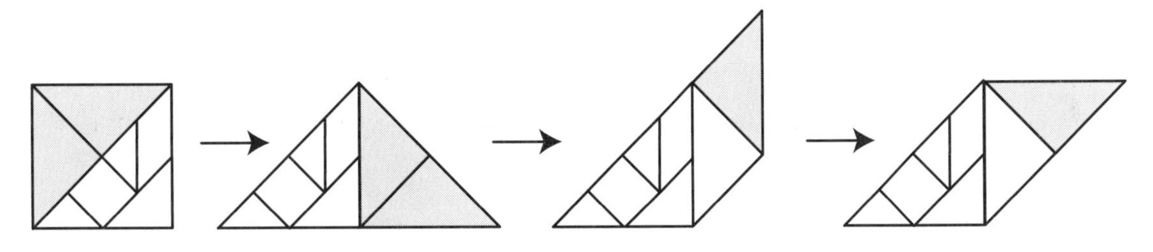

By using the piece of spaghetti to test for concavity or convexity, students reinforce their understanding of these concepts and strengthen their spatial visualization skills. Students might also be encouraged to investigate the measures of the angles within their polygons. Some students may recognize that in order for a polygon to be convex, each of its interior angles must measure less than 180°.

Students should discover that only 13 unique convex polygons can be formed from the 7-piece Tangram set. Different arrangements of the same 7 pieces may yield congruent shapes. To check for congruence, students may reflect, flip, and/or rotate one shape to see whether it matches another shape. The 13 solutions are shown here and on the next page, each with one possible arrangement of the 7 pieces from the Tangram set.

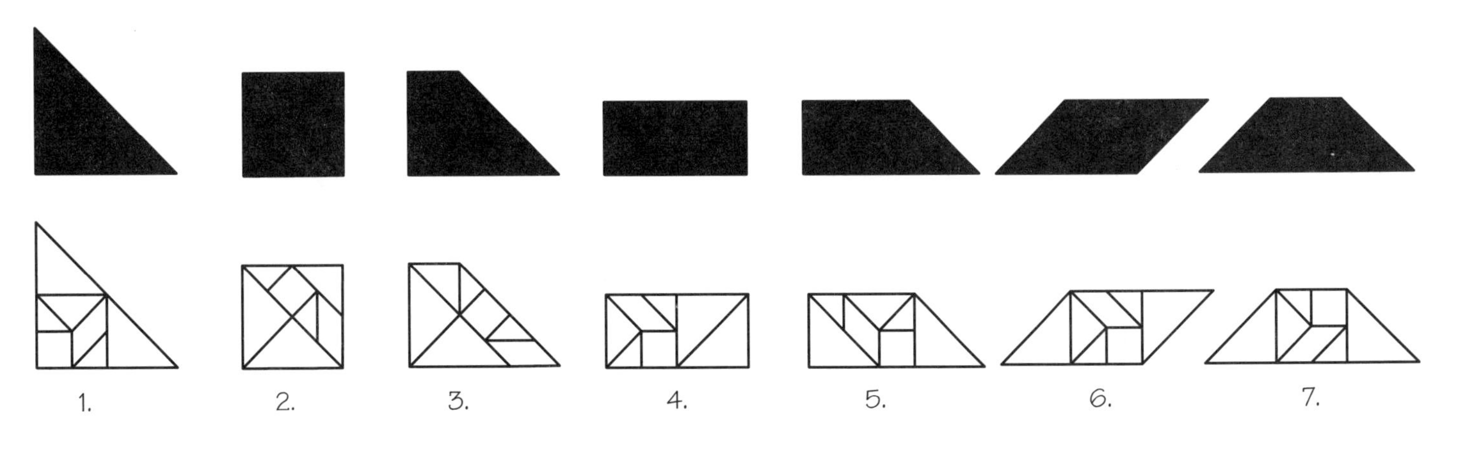

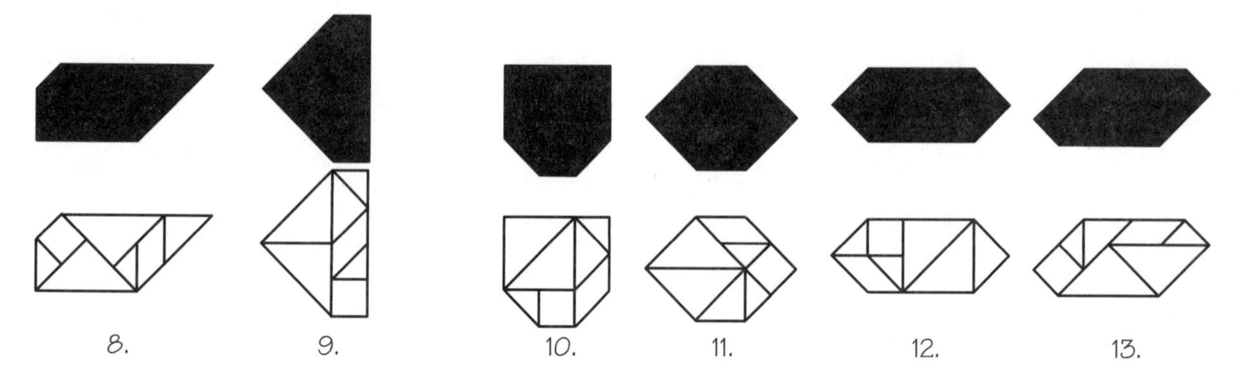

Students may note that there are more convex polygons with an even number of sides than there are with an odd number of sides. They may also point out that there are many more concave polygons that can be made than convex polygons.

When the two Tangram sets are used together to generate convex circuit boards, students will discover that there are many more solutions possible. Rather than starting with 14 separate Tangram pieces, students may choose to combine two existing convex shapes from the first activity to generate other polygon shapes that can then be tested for convexity or concavity. Several examples of these new convex circuit boards are shown below. The shapes will still have at least 3 sides; however, students will discover that with two sets, it is possible to design convex circuit boards that have more than 6 sides.

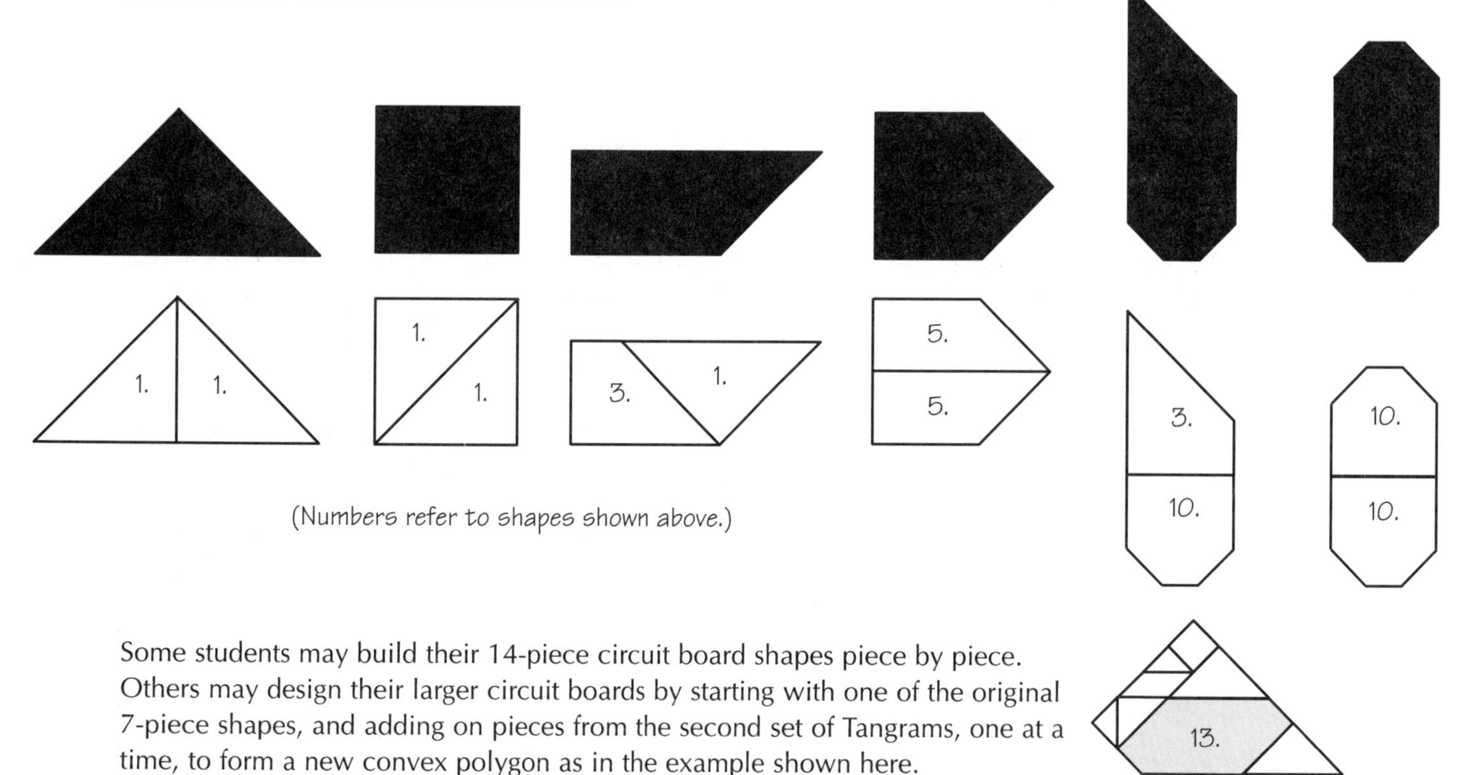

(Numbers refer to shapes shown above.)

Some students may build their 14-piece circuit board shapes piece by piece. Others may design their larger circuit boards by starting with one of the original 7-piece shapes, and adding on pieces from the second set of Tangrams, one at a time, to form a new convex polygon as in the example shown here.

Students may be interested to discover that many of the new convex polygons made from two sets of Tangram pieces are similar to those made from one set. You may want to use this discovery to encourage students to search for other pairs of similar polygons that can be made from sets of Tangrams.

STAR SEARCH

- Properties of geometric figures
- Angle measures of polygons
- Looking for patterns
- Spatial visualization

Getting Ready

What You'll Need

Circular Geoboards, 1 per student

Rubber bands

Circular geodot paper, page 110

Scissors

Activity Master, page 98

Overview

Students create polygons on the circular Geoboard and investigate patterns formed by their diagonals. In this activity, students have the opportunity to:

- understand what a diagonal is
- see that diagonals from a single vertex in a polygon partition the polygon into triangles
- investigate relationships between the number of sides of the polygon and the sum of the measures if its interior angles
- determine the total number of diagonals in a polygon

Other *Super Source* activities that explore these and related concepts are:

Geoboard Challenge, page 39
Braille Puzzles, page 43
Circuit Boards, page 47

The Activity

On Their Own (Part 1)

Islamic art is known for its rich, intricate geometric patterns, all of which are based on the circle. In the Islamic culture, the artist and the mathematician are one and the same person. The art uses underlying square or triangular pattern grids to create mosaic designs. Some designs are based on diagonals of polygons. What can you discover about these diagonals?

- Work with a partner. Each of you should make a polygon with a different number of sides on the circular side of your Geoboard. The vertices of your polygons should be located at pegs on the circle, not at the center peg.
- Select any vertex of your polygon and make as many diagonals as possible from that vertex.
- Draw your polygon and diagonals on circular geodot paper. Record the number of sides and vertices in the polygon, the number of diagonals drawn from one vertex, and the number of triangles created in the interior.
- Repeat the process for polygons with different numbers of sides. Try at least 6 different examples. Be sure to record your polygons and data.
- Cut out your geodot drawings and organize them with your partner's drawings. Look for patterns and be ready to talk about what you noticed.

Thinking and Sharing

Divide the chalkboard into columns with headings *3 sides, 4 sides, 5 sides, ..., 12 sides.* Ask a student who has made a 3-sided polygon to post his or her drawing and data in the first column. Invite other students with different 3-sided polygons to add theirs to the column. Continue posting the polygons in this way under the other column headings. Have students compare their findings with those of other pairs.

Use prompts like these to promote class discussion:

- What do you notice when you look at the polygons?
- What is the least number of diagonals a polygon has? the greatest number?
- Were you ever able to predict the number of diagonals that could be drawn from a vertex before making them? If so, how?
- How do the number of sides and vertices of a polygon compare?
- How did the number of diagonals from a vertex compare to the number of sides in the polygon? How many diagonals would there be from a vertex of a polygon with n sides?
- How is the number of triangles in the partitioning related to the number of sides of the polygon? How many triangles would there be in a polygon with n sides?
- How could you find the sum of the measures of the angles in any of these polygons knowing the sum of the measures of the angles in a triangle?

On Their Own (Part 2)

What if... *the Islamic artisan/mathematician wanted to use diagonals to create stars with different numbers of points? How many diagonals would be needed to create these stars in polygons with different numbers of sides?*

- From your work in Part 1, select any polygon with 5 or more sides and recreate it on your Geoboard. Make as many diagonals as you can from each vertex of the polygon.
- Draw your polygon and diagonals on circular geodot paper. Record the number of vertices in the polygon, the number of diagonals drawn from each vertex, and the total number of diagonals in the polygon.
- Repeat the process using other polygons from the first activity. Try at least 6 different examples. Be sure to record your polygons and data.
- Look for patterns in your data. Also look to see whether you think some of your stars are esthetically more pleasing than others. Be ready to talk about your observations.

Thinking and Sharing

Using the chart created for the first activity, ask students to complete the diagrams of the polygons they worked on, drawing in the remaining diagonals from each vertex. Have them also indicate the total number of diagonals contained in the polygon.

Use prompts like these to promote class discussion:

- What did you notice about the different polygons and the stars they produce?

- Which star(s) do you find most pleasing? Why?
- Which star(s) do you find most unusual? What is it about the polygon(s) containing those diagonals that made such an unusual star?
- What patterns did you discover in your data?
- Why is the total number of diagonals not the product of the number of diagonals from one vertex and the number of vertices?
- How many diagonals do you think could be drawn in a polygon with 13 sides? Explain your reasoning.
- How many diagonals can be drawn in a polygon with *n* sides? Explain.

Picture a 20-sided polygon. Figure out the number of diagonals that could be drawn from one of the vertices of the polygon, the number of triangles that would be formed by these diagonals, and the total number of diagonals that could be drawn in the polygon. Then write a convincing argument supporting your conclusions.

Teacher Talk

Where's the Mathematics?

The polygons that can be made on the Geoboard range in size from those with 3 sides (and 3 vertices) to those with 12 sides (and 12 vertices). As solutions are posted on the class chart, students may be surprised at the variety of shapes that can be made by selecting different pegs on the Geoboard for the vertices. They may also be surprised at how different the shapes look, depending on the choice of vertex used to form the diagonals. Examples are shown below:

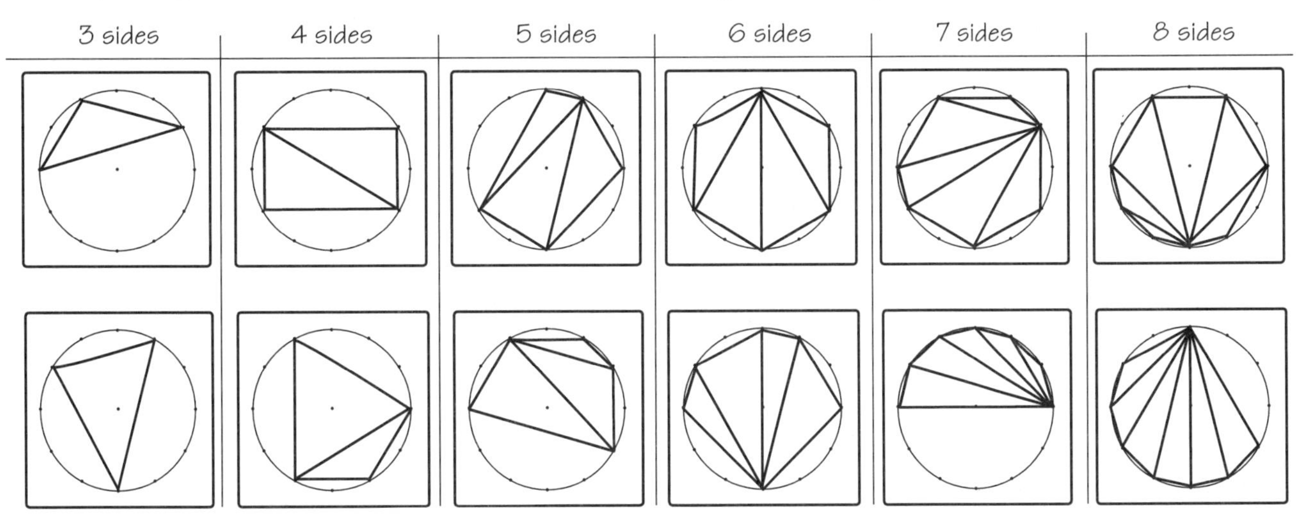

From their data, students may notice that the number of diagonals drawn from any vertex of the polygon is always 3 less than the number of sides or vertices of that polygon. Once a vertex is chosen, diagonals can be drawn by connecting that vertex to all but 3 vertices – itself, and the 2 adjacent vertices of the polygon. A 3-sided polygon will, therefore, have no diagonals; a 4-sided polygon will have 1 diagonal from each vertex; a 5-sided polygon will have 2 diagonals from each vertex, and so on. In general, if *n* represents the number of sides or vertices of a polygon, the expression $(n - 3)$ represents the number of diagonals that can be drawn from each vertex of the *n*-sided polygon.

Students may also recognize that as the diagonals are created, the number of triangles formed is always 2 less than the number of sides or vertices of the polygon. Thus, if n represents the number of sides or vertices of a polygon, the expression $(n - 2)$ represents the number of triangles into which the figure can be partitioned. Students may also discover that the number of triangles is always one more than the number of diagonals.

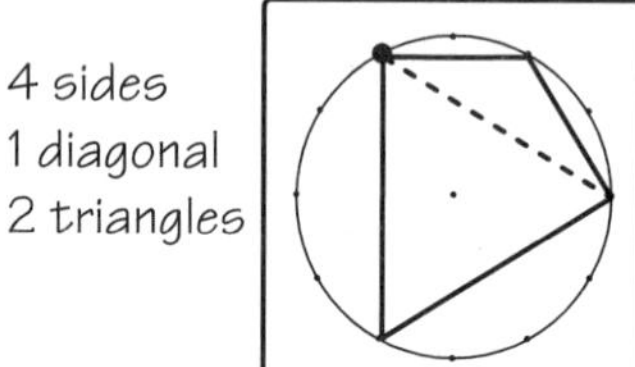

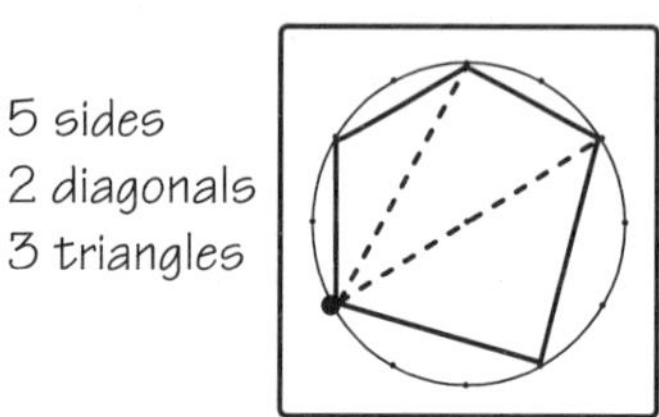

6 sides
3 diagonals
4 triangles

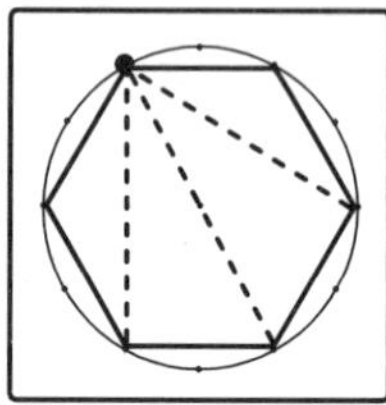

Knowing that the sum of the measures of the angles of a triangle is 180° enables students to find the sum of the interior angle measures of any polygon. Since the angles of the polygon are composed of the angles of the triangles formed by the diagonals, the sum of the interior angles can be calculated by multiplying the number of triangles in the partitioning by 180°. The algebraic expression representing this relationship is $180° \times (n - 2)$, where n represents the number of sides or vertices in the original polygon.

In Part 2, as students consider the total number of diagonals in their polygons, they will discover that the same number of diagonals originate from each vertex. With this fact in mind, the strategy of multiplying the number of diagonals originating from each vertex by the number of vertices (or sides) of the polygon is helpful in determining the total number of diagonals. However, this product must then be divided by 2 in order to adjust for having counted each diagonal twice. In general, if n represents the number of sides or vertices of a polygon, the expression $\frac{(n)(n-3)}{2}$ represents the total number of diagonals.

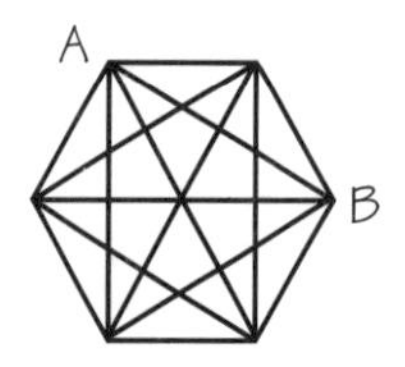

Diagonal $\overline{AB}$ is the same as diagonal $\overline{BA}$, and should be counted only once.

6 vertices, 3 diagonals from each vertex

$$\frac{6 \times 3}{2} = 9 \text{ diagonals}$$

Some students may decide that the stars created from diagonals of equilateral polygons are esthetically more pleasing because of their symmetry. Students may also suggest that the more unusual-looking stars are those formed by the diagonals of polygons that contain an angle or angles that are either very small or very large. Students may be interested in further analyzing the attributes of polygons that affect the shapes of the stars formed by the diagonals.

Investigating Angles

1. Spider Web Site, page 56 (Circular Geoboards)
2. Patangles, page 60 (Pattern Blocks)
3. M.C. and Me, page 64 (Pattern Blocks)

The activities in this cluster give students a wide range of practice working with angles, making these good opportunities for using and reinforcing angle vocabulary. Material presented as new in these activities includes inscribed angles, intercepted arcs, and exterior angles. These terms are defined for the student in the *On Their Own* features.

1. Spider Web Site *(Investigating angle measure and reinforcing angle vocabulary)*

In this activity students relate the measures of inscribed angles to the measures of the intercepted arcs. The activity provides a good opportunity to reinforce vocabulary about angle classification. During the activity students may form conclusions about interior angle sums of polygons as well as about properties of triangles. Some of these conclusions are discussed on page 59 in *Where's the Mathematics?*

The terms "inscribed angle" and "intercepted arc" are explained in *On Their Own* Part 1 and do not need to be prereviewed.

2. Patangles *(Exploring angle relationships)*

In this activity students examine interior and exterior angles of polygons, and in the process of organizing their data they make discoveries about angle relationships.

On Their Own Part 1 assumes that students can recognize an interior angle; the activity does, however, explain how to recognize an exterior angle. Part 1 also defines "angle bisector." If teachers want to clarify these terms for students before the activity, they could use the diagrams given at the beginning of *Where's the Mathematics?*, page 62.

Where's the Mathematics? also discusses (on page 63) some of the properties students may discover, including exterior angle sums and the supplementary relationship of exterior-interior angle pairs.

3. M.C. and Me *(Exploring tessellation)*

In this activity students explore ways of surrounding a point in a plane with various polygons. The activity would work well as an introduction to tessellation.

On Their Own Part 1 introduces the idea of tessellation first locally (surrounding a point) and then generally (repeating the pattern to tile the plane). It assumes that students know the term "tessellation," so teachers should be sure to introduce the idea before beginning the activity. One good way to introduce the idea would be to show examples of Escher designs. After showing these designs, teachers could give a simple tessellation example made from squares: Each vertex consists of four right angles that combine to surround the point, and the design can be extended to cover the plane.

Where's the Mathematics? contains numerous examples of tessellations, some mathematical mosaics (the same arrangement of shapes surrounding each vertex) and some not. Photocopies of some of these might help spur class discussion.

SPIDER WEB SITE

- Inscribed polygons
- Inscribed angles
- Interior angles of polygons
- Congruence

Getting Ready

What You'll Need

Circular Geoboard

Rubber bands

Rulers

Circular geodot paper, page 110

Activity Master, page 99

Overview

Students investigate the angle measures of polygons that can be inscribed in a circular Geoboard. In this activity, students have the opportunity to:

- learn that the measure of an inscribed angle is half the measure of its intercepted arc
- find the measures of the interior angles of various inscribed polygons
- investigate the sum of the measures of the interior angles of polygons
- use logical reasoning to search for all possibilities

Other *Super Source* activities that explore these and related concepts are:

Patangles, page 60

M.C. and Me, page 64

The Activity

On Their Own (Part 1)

Most children learn the song about the "itsy-bitsy spider" that goes up the water spout. Suppose the spider decided to build a web inside the water spout. If it must start by weaving a triangular web whose vertices touch the inner walls of the circular spout, can you help the "itsy-bitsy" spider figure out the number of ways it can start weaving its web?

- Use a circular Geoboard to represent the inside of the water spout. Working with a partner, create an inscribed triangular "web" on your Geoboard. Remember that in inscribed triangles, all 3 vertices must lie on the circle.

Inscribed triangle

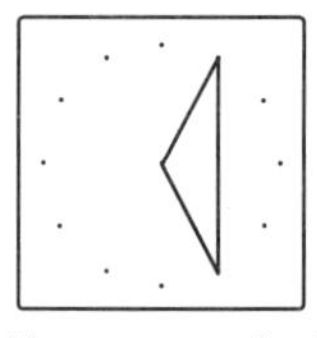

Not an inscribed triangle

- Find the measure of each angle in your triangular web. To do this, remember that:
  - A circle contains 360°.
  - The measure of an inscribed angle is half the measure of the arc it intercepts.

intercepted arc

inscribed angle

- Using a ruler, record your triangle on circular geodot paper. Label each angle with its measure.

- Find as many different inscribed triangular webs as you can and record your findings.
- Organize your work and look for patterns.

Thinking and Sharing

Ask one pair of students to recreate one of the triangles they found on their Geoboard, display it on the chalk rail, and write the measures of its angles above it. Continue with other pairs until students are confident all possible triangles have been displayed.

Use prompts like these to promote class discussion:

- How many different triangular webs did you find?
- Did you use a strategy in creating new triangles? If so, explain what you did.
- How did you find the measures of the intercepted arcs?
- What methods did you use to decide if all the triangles were different?
- How were you sure that all possible triangles were found?
- What number patterns did you discover?
- What relationship exists between the number of degrees in the inscribed angles of a triangle and the number of degrees in a circle? How can you explain this?

On Their Own (Part 2)

What if... ***the "itsy-bitsy" spider decided to try different polygon shapes for the beginning of its web? How can you use what you know about inscribed angles to find the measures of the angles in its new web?***

- Decide with your partner whether the new polygon web will contain 4, 5, or 6 sides.
- Create an inscribed polygon with the agreed upon number of sides on your circular Geoboard. Figure out the number of degrees in each inscribed angle of the polygon web.
- Using a ruler, record the inscribed polygon on circular geodot paper. Label each angle with its measure.
- Find as many different inscribed webs as you can having the given number of sides. Record your findings.
- Organize your work and look for patterns.

Thinking and Sharing

Group pairs according to the number of sides they have selected for their inscribed polygon webs — 4, 5, or 6. Have the students within each group share their work and prepare a display of their findings.

Use prompts like these to promote class discussion:

- How many different 4-sided (5-sided, 6-sided) polygon webs can be made?

- How was what you learned in the first part of the activity helpful in this part?
- Did you use a strategy in creating new polygons? If so, explain what you did.
- What methods did you use to decide whether all of your polygons were different?
- How were you sure you had found all possible polygons?
- What number patterns did you discover?
- What relationships exist between the interior angle measures of 3-, 4-, 5-, and 6-sided figures?

Write an explanation of how the sum of the measures of the angles of a triangle is related to the number of degrees in a circle. Use words such as inscribed polygon, vertex, inscribed angle, intercepted arc, angle measure, etc. in your explanation.

Teacher Talk

Where's the Mathematics?

In this activity, students apply the fact that the measure of an inscribed angle is half the measure of its intercepted arc to discover many more facts. Students should find that there are 12 different inscribed triangles that can be constructed on the circular Geoboard.

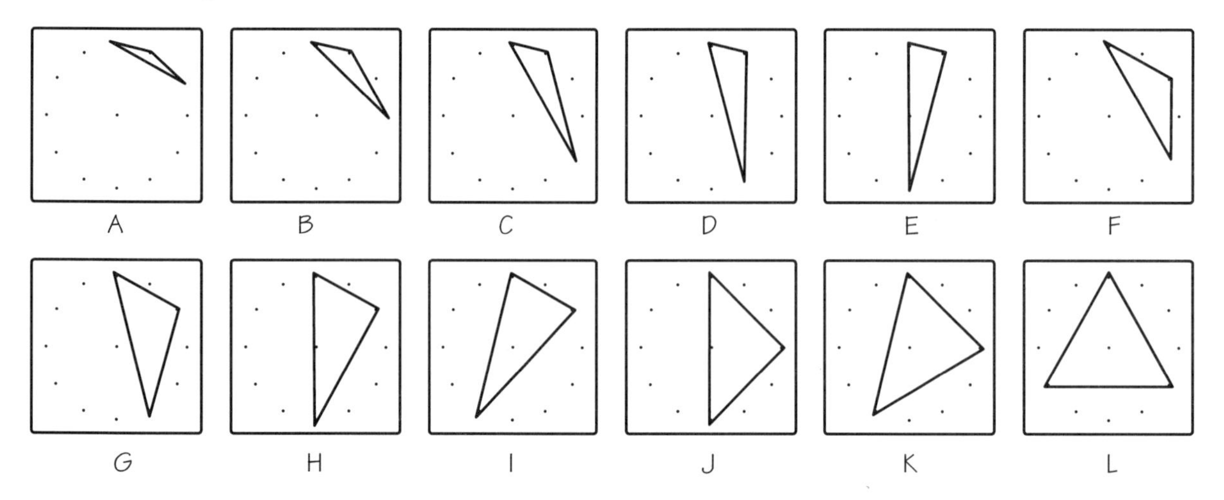

Many students may already be familiar with the fact that the sum of the measures of the angles of a triangle is 180°. Using the relationship between the inscribed triangles and the circular Geoboard helps confirm this powerful concept. Students may point out that the three angles of the triangle intercept arcs that together make up the entire circle. Since the arc making up the entire circle measures 360° and the measure of an inscribed angle is half the measure of its intercepted arc, the sum of the three angles of the triangle is half of 360°, or 180°.

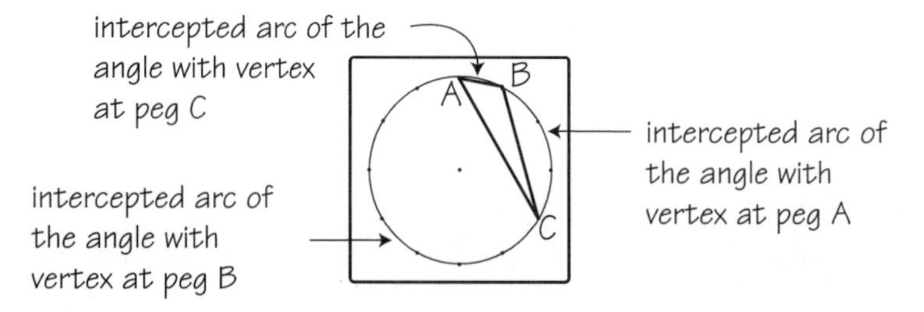

As students organize their data, they may notice that each angle measure is a multiple of 15°. Since the smallest intercepted arc has a measure of 30°, (360°÷ 12), the smallest angle measure in any of these inscribed triangles will be 15°. Each time the side of any inscribed angle is moved to an adjacent peg on the Geoboard to form a larger arc, another 30° is added to the measure of the intercepted arc, increasing the measure of its inscribed angle by 15°.

If students choose to organize their data in a chart, they may find it helpful to start with the triangles whose smallest angle measures 15°, then those triangles whose smallest angle measures 30°, then 45°, and then 60°, as shown in the chart below.

	first angle	second angle	third angle		first angle	second angle	third angle
A	15	15	150	G	30	45	105
B	15	30	135	H	30	60	90
C	15	45	120	I	30	75	75
D	15	60	105	J	45	45	90
E	15	75	90	K	45	60	75
F	30	30	120	L	60	60	60

NOTE: Letters refer to figures on previous page.

By looking at the data, students may notice that

- the sum of the measures of the angles of any triangle is 180°
- a triangle cannot contain more than one right or one obtuse angle
- when a triangle contains a 90° (right) angle, the remaining two angles have measures that total 90°

Some students may organize their triangles according to angle types. There are 3 acute triangles, 3 right triangles, and 6 obtuse triangles. Each of the right triangles has a diameter of the circle for its longest side (hypotenuse). All of the obtuse triangles lie on one side of an imaginary diameter while the acute triangles cross the imaginary diameter at 2 points.

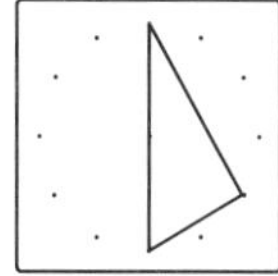
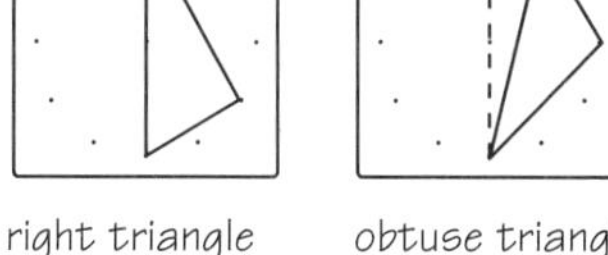
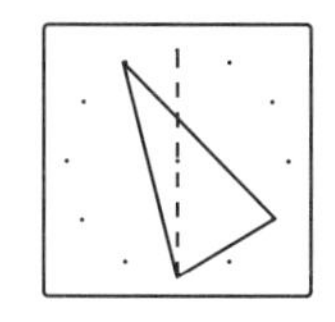

right triangle obtuse triangle acute triangle

Students may have used one or more of the following methods to determine whether they found all possible triangular webs:

- systematically moved rubber bands from peg to peg on the Geoboard
- organized their data in a specific way
- checked solutions against those of other students

Similar methods can be used to create all 4-, 5-, or 6-sided inscribed polygon webs, check for duplications, and organize the data about their angle measures. Students should discover that the sum of the angle measures of 4-sided polygons is 360°; of 5-sided polygons is 540°; and of 6-sided polygons is 720°. Students may notice that as the number of sides of a polygon increases by one, the sum of its interior angle measures increases by 180°.

Examples of several different 4-, 5-, and 6-sided inscribed polygon webs are shown at right.

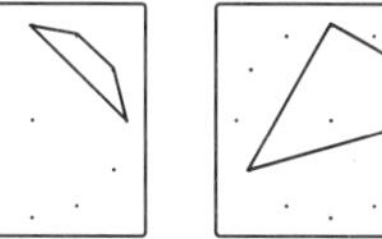
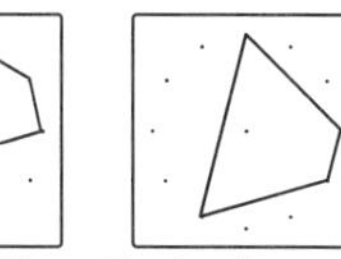
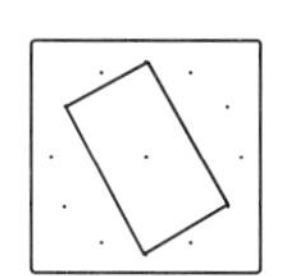

4-sided inscribed polygons

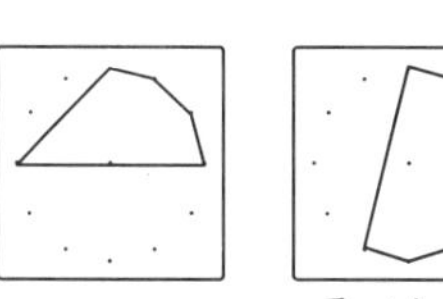
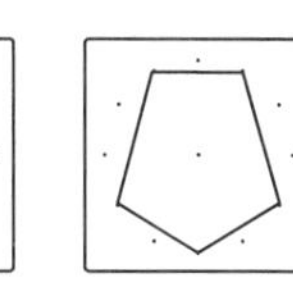

5-sided inscribed polygons

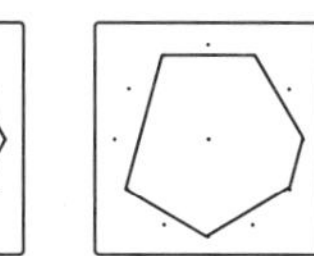
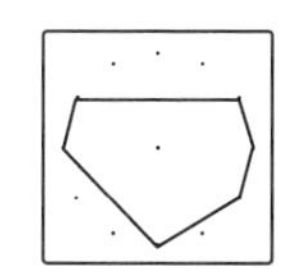

6-sided inscribed polygons

PATANGLES

- Angle measures
- Interior and exterior angles of polygons
- Angle bisectors
- Supplementary angles

Getting Ready

What You'll Need

Pattern Blocks, several of each shape per pair

Rulers

Activity Master, page 100

Overview

Students investigate the measures of the interior and exterior angles of the Pattern Block shapes, and the measures of the angles formed by bisecting these angles. Students then build as many new angles as they can using these angles. In this activity, students have the opportunity to:

- identify and measure interior and exterior angles of polygons
- use spatial reasoning to find angle measures
- understand the concept of bisecting an angle
- combine angles using addition, subtraction, or a combination

Other *Super Source* activities that explore these and related concepts are:

Spider Web Site, page 56

M.C. and Me, page 64

The Activity

On Their Own (Part 1)

Pat Bangle loves to build new angles (which he calls "Patangles") based on combinations of angle measures found in the 6 Pattern Block shapes. In order to create these "Patangles," Pat needs to know the measures of the angles he has to work with. Can you find these angle measures?

- Working with your partner, find the measure of each interior angle of each Pattern Block shape. To get started, remember that the green triangle has 3 congruent angles, and that the sum of the angles in any triangle is 180°.
- Trace each Pattern Block shape on a piece of white paper and record its angle measures. Also record the number of angles in the polygon.
- Using a ruler, extend one side of each polygon at each vertex. These angles formed outside the polygon are called exterior angles. Use your Pattern Blocks to find the measure of each exterior angle of each Pattern Block shape. Remember that the measure of a straight angle is 180°.

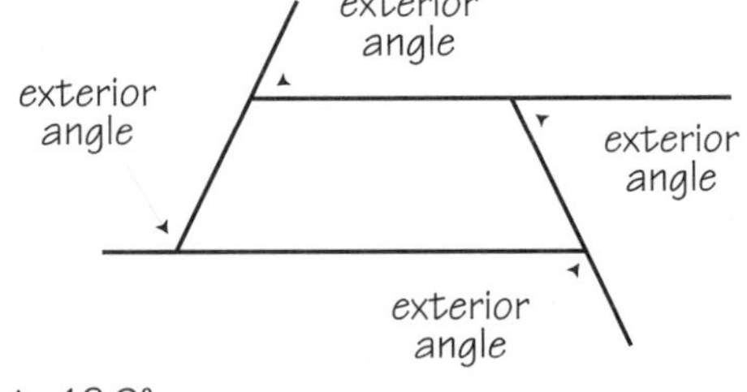

- Record the measure of each exterior angle of the Pattern Block shapes.
- Bisecting an angle means dividing it into 2 congruent angles. Use paper-folding to bisect each of the interior and exterior angles of each Pattern Block shape.
- Label each new angle with its measure. Look for angle measures that you have not already found.
- Organize your data in a chart. Look for patterns in your data.

Thinking and Sharing

Have students help you complete a chart with the following headings: *Pattern Block Shape, # of angles, measures of interior angles, measures of exterior angles, measures of angles formed by bisecting angles.*

Use prompts like these to promote class discussion:

- How did you find the measures of the interior angles of each shape?
- What did you notice about the interior angle measures?
- How did knowing the fact that the measure of a straight angle is 180° help in finding the measure of the exterior angles?
- What did you notice about the exterior angle measures?
- Did anyone notice anything about the sum of the measures of the interior angles? of the exterior angles? If so, tell what you noticed.
- What other patterns did you find?
- What did you notice about the measures of the angles you could form by bisecting existing angles?

On Their Own (Part 2)

What if... *Pat Bangle invited you to help him build Patangles? What Patangles with measures between 0° and 360° can you create?*

- Working with your partner, build a new angle based on combinations of the Pattern Block angle measures. Here's how:
  - First look at the interior and exterior angles you found in the first activity.
  - Use addition, subtraction, or a combination of both on these angles to create a new angle (a Patangle).
  - Your Patangle should have a measure that is different from that of any angle you have found so far.
- Trace your Patangle on a piece of paper, showing the angles used to build it. Record its measure and the measure of the angles you used to build it.
- Find as many Patangles as you can. Record each one as described above.
- Be ready to discuss any patterns in the measures of your Patangles.

Thinking and Sharing

Invite students to share their Patangles with the class. Make a list of all new angle measures generated by the Patangles.

Use prompts like these to promote class discussion:

- How many different Patangles were you able to create?
- What methods did you use to build the new angles?
- Did you ever find more than one way to build a particular Patangle? If so, give an example.
- What was hard about building Patangles? What was easy?
- What patterns did you find?
- Do you think we've found all possible Patangles? How do you know?
- Can you think of any angle measures that cannot be generated using the Pattern Blocks? If so, give some examples.

Write a letter to Pat Bangle describing the method(s) you used to create Patangles. Include any diagrams that might be helpful in illustrating your methods.

Teacher Talk

Where's the Mathematics?

Students work with three types of angles in this activity: interior angles of polygons, exterior angles of polygons, and bisected angles.

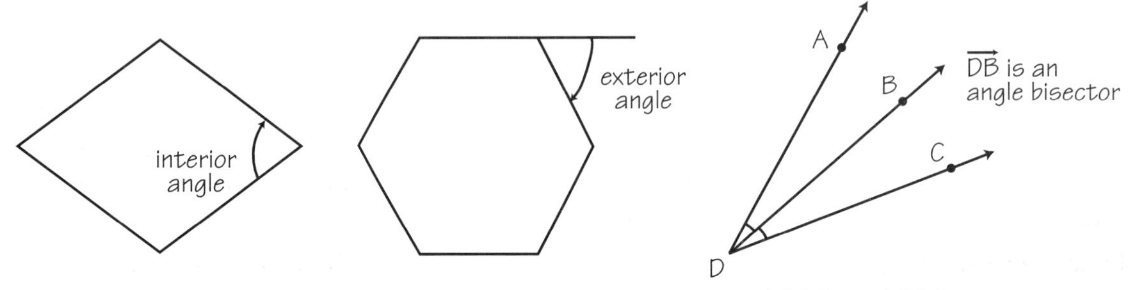

Students should be able to identify the different types of angles on their diagrams. Encourage them also to use the terms during the discussion.

It is likely that students will recognize the measures of the interior angles of the square to be 90°. Using the fact that the triangle is equiangular, students can determine that each interior angle measures 60°. They can then use these angles to find the measures of the interior angles of the other four Pattern Block shapes. For example, students can find that each interior angle of the hexagon is equivalent to two angles of the triangle; thus, each measures 120°. Since two of the smaller angles of the tan rhombus fit inside one angle of the triangle, the measure of the smaller angle of the tan rhombus is 30°. Similar methods can be used to find the measures of the other interior angles.

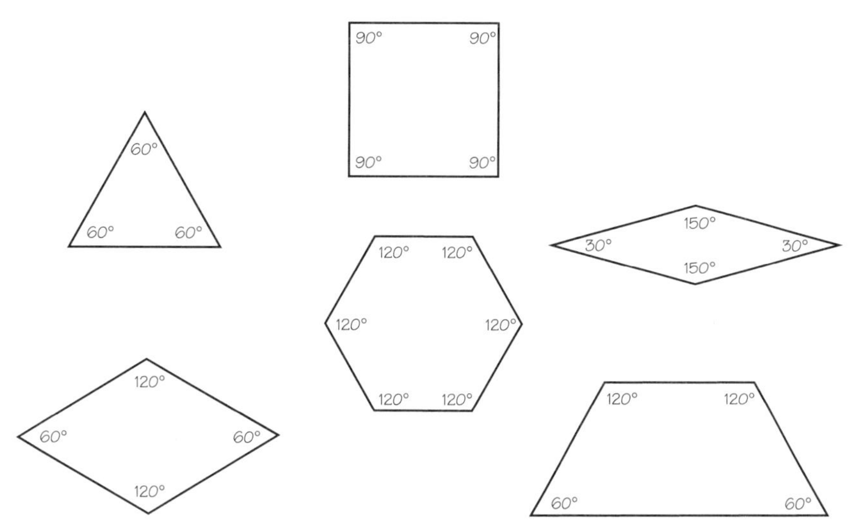

Students should observe that each interior angle forms a straight angle with its adjacent exterior angle. Therefore, their angle measures total 180°. This discovery provides a nice opportunity to introduce the concept of supplementary angles (two angles whose measures total 180°).

Some students may notice that the sum of the measures of the exterior angles of each polygon is 360°. Those who look at the sums of the interior angles may notice that for the four quadrilaterals, the sum of the interior angles is 360°. Students may also point out that all of the interior and exterior angles of the Pattern Blocks are multiples of 30°. Bisecting the 30°, 90°, and 150° angles produces three additional angle measures to work with: 15°, 45°, and 75°.

Pattern Block shape	# of angles	interior angles	exterior angles	angles formed by bisecting
triangles	3	60°	120°	(30°, 60°)
squares	4	90°	90°	45°
tan rhombus	4	30°, 150°	30°, 150°	15°, 75°
blue parallelogram	4	60°, 120°	60°, 120°	(30°, 60°)
trapezoid	4	60°, 120°	60°, 120°	(30°, 60°)
hexagon	6	120°	60°	(30°, 60°)

Thus, the angle measures that are available for creating Patangles are 15°, 30°, 45°, 60°, 75°, 90°, 120°, and 150°, all multiples of 15°. As students begin building their Patangles, they may discover that all new angles that can be built will also be multiples of 15°. Thus, any angle that is not a multiple of 15° cannot be formed using the original set of angles.

To build their Patangles, students may trace the angles from either their actual Pattern Blocks or from their recordings. They may form new angles using addition of existing angles (as in figure 1), subtraction of existing angles (as in figure 2), or a combination of both (as in figure 3). Students may also enjoy investigating the different ways of forming a particular angle.

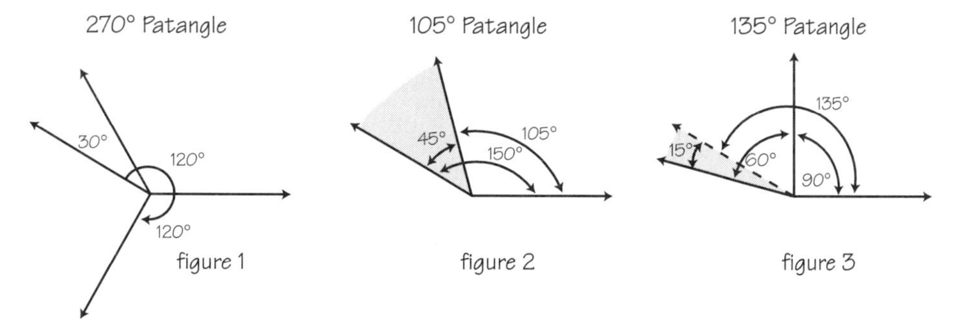

figure 1 figure 2 figure 3

M.C. AND ME

- Tessellations
- Angles in polygons
- Spatial visualization

Getting Ready

What You'll Need

Pattern Blocks, 1 set per pair

Colored pencils or markers

Activity Master, page 101

Overview

Students explore ways to arrange Pattern Blocks around a point to create tessellating designs. They investigate the angles of the shapes in their designs and explore combinations of shapes that can be used to create mathematical mosaics. In this activity, students have the opportunity to:

- use spatial reasoning to find angle measures
- devise strategies for testing shapes to determine whether or not they tessellate
- develop spatial visualization skills
- analyze a pattern to see if it is a mathematical mosaic

Other *Super Source* activities that explore these and related concepts are:

Spider Web Site, page 56

Patangles, page 60

The Activity

On Their Own (Part 1)

M.C. Escher, the noted Dutch artist who lived from 1898 to 1972, often visited the Alhambra in Spain to gain inspiration for his work. Many of Escher's tessellating creatures and geometric shapes were based on the magnificent tile patterns he found while visiting there. How can you use Pattern Blocks to create tessellating designs?

- Working with your partner, find the measure of each angle of each Pattern Block shape. To get started, remember that the green triangle has 3 congruent angles, and that the sum of the angles in any triangle is 180°.
- Trace each Pattern Block shape on paper and record its angle measures.
- Draw a point in the center of a clean sheet of paper. Using only green triangles and placing them so that a vertex of each triangle touches your point, determine the number of triangles needed to surround the point completely.
- Trace each triangle as it appears in the tessellating pattern and determine the sum of the angle measures surrounding the point. Record this sum and the number of triangles needed to surround the point.

- Repeat the process with the other five Pattern Block shapes. For some shapes there may be more than one way to arrange them. Record your different arrangements and look for patterns in your data.
- Select your favorite tessellating arrangement, and explore ways to connect several identical arrangements together to form a larger tessellating design. Remember that in a tessellation, there can be no spaces or gaps between the shapes.

Thinking and Sharing

Beginning with their triangle arrangements, have students post their drawings along with their data about the sum of the angle measures and the number of Pattern Blocks needed to surround a point completely. Do the same for each of the Pattern Block shapes. Then invite students to walk around the room and examine the tessellations made by their classmates.

Use prompts like these to promote class discussion:

- How did you find the measures of the angles of each Pattern Block shape?
- Which Pattern Block shapes are equiangular?
- How many triangles (squares, trapezoids, etc.) were needed to surround the point in each arrangement?
- For which Pattern Block shapes was there only one way to surround the point? For which was there more than one way? How can you explain this?
- What patterns did you discover?
- What discoveries did you make when creating your larger tessellating design?

On Their Own (Part 2)

What if... ***you wanted to build tessellations of the type described as "mathematical mosaics?" If a mathematical mosaic is a design that has the same arrangement of shapes surrounding every vertex in the design, what mathematical mosaics can you create using Pattern Blocks?***

- Working with your partner, select 3 or more Pattern Block shapes that you might like to use in your mathematical mosaic.
- Draw a point in the middle of a clean sheet of paper and try to create an arrangement that completely surrounds your point using the shapes you selected. If you are unable to complete the tessellation around the point, change one or more of the Pattern Block shapes until you obtain a tessellation.
- Explore ways to extend your tessellation to form a larger design made from the same shapes. In order for your design to be a mathematical mosaic, the arrangement of shapes surrounding every vertex in your design must be exactly the same.
- Experiment with building mathematical mosaics using other combinations of Pattern Block shapes. See what you can discover about the shapes and designs that can be used to form mathematical mosaics.

Thinking and Sharing

Invite students to walk around the room and examine the designs made by their classmates.

Use prompts like these to promote class discussion:

- How did you decide what shapes to use for your mathematical mosaics?
- Did you use what you know about the angle measures of the Pattern Blocks to help in selecting the shapes? If so, explain.
- How did you determine the arrangement of the shapes around the given point?
- How did you go about extending your initial pattern?
- How did you check to see whether your design was a mathematical mosaic?
- What shapes were easy to use in building mathematical mosaics? What shapes were difficult to use? Why?
- What do you notice about the shapes and designs that form mathematical mosaics?

Students may be interested in recording their mathematical mosaics. You might have them either trace the shapes onto paper and color them, or use construction-paper cutouts of the various Pattern Block shapes and paste them onto a sheet of construction paper.

Find five tessellating patterns in your home or environment. Describe the types of polygons used in the designs and their angle measures. Decide whether the tessellating patterns are mathematical mosaics and support your reasoning. Include a sketch of each pattern.

Teacher Talk

Where's the Mathematics?

Working with tessellations provides students with an opportunity to learn about angle measures and to see how the different Pattern Block shapes are related to each other. Students may recognize patterns that appear on floors and walls at home and in school as being tessellations, patterns of non-overlapping shapes that completely cover a surface without leaving any spaces or gaps between the shapes.

To find the measures of the angles of the Pattern Block shapes, students will need to compare the angles and use their understanding of equivalence. Using the fact that the triangle is equiangular, students can determine that the measure of each of its angles is 60°. This angle can then be used to help determine the angle measures of the other blocks. For example, since the smaller angle of the blue parallelogram is equivalent to the angle of the triangle, and the larger angle of the blue parallelogram is equivalent to two angles of the triangle, their measures are, respectively, 60° and 120°. Similar equivalences exist in the trapezoid and hexagon.

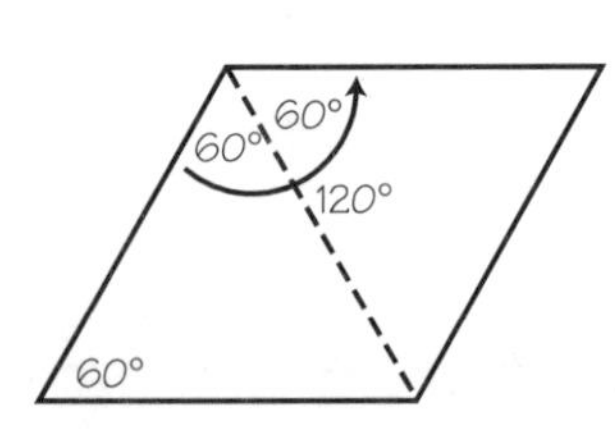

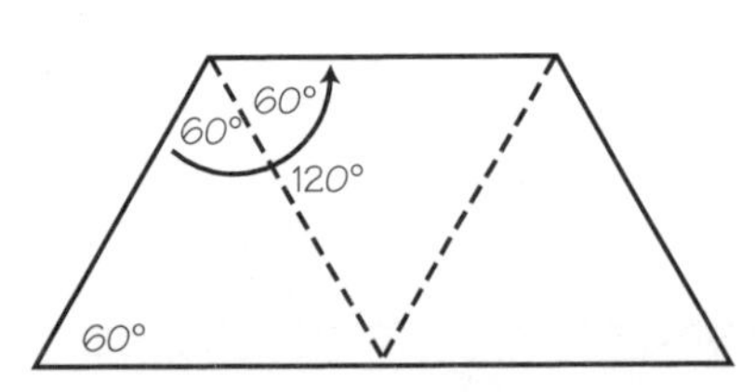

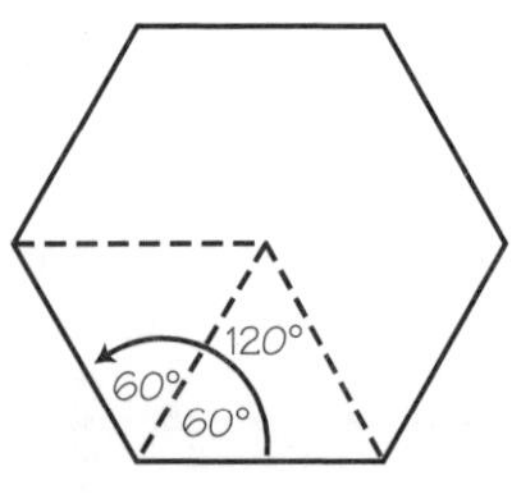

Since two of the smaller angles of the tan rhombus fit within one angle of the triangle, the measure of the smaller angle of the tan rhombus is 30°. This angle can then be used to determine that the measure of the larger angle in the tan rhombus is 150° and to verify that the angles of the square are, indeed, 90°.

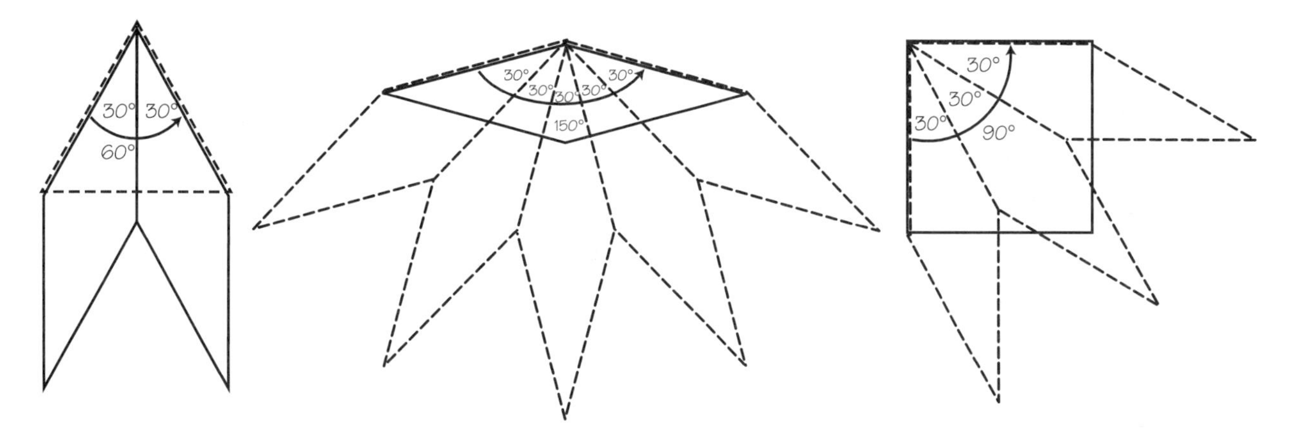

The triangle, square, and hexagon are the only regular (both equilateral and equiangular) Pattern Blocks, and they will each surround a given point in only one way. Students should notice that in each case, the sum of the angles surrounding the point is 360°.

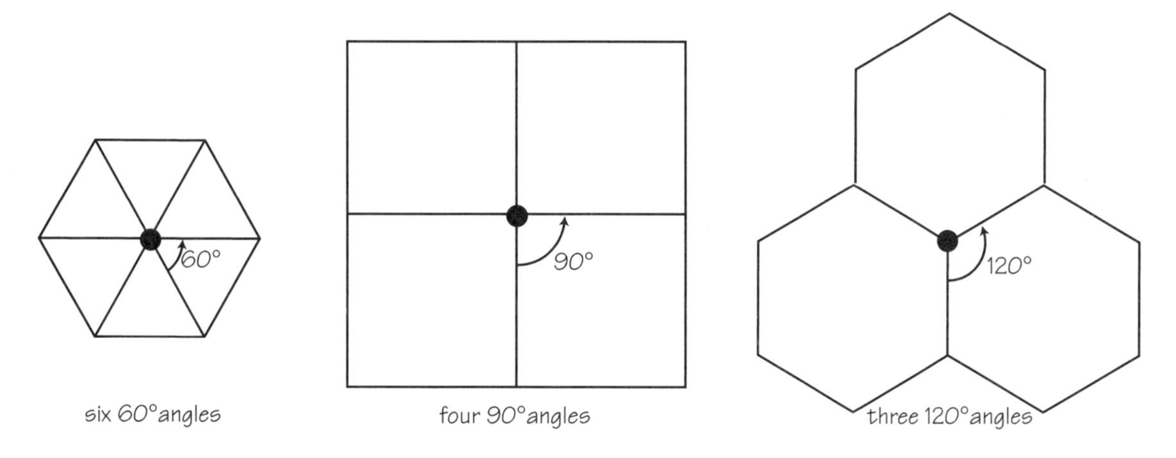

The rhombus and trapezoid shapes are not equiangular, so they can be arranged around the point in different ways. Students should be encouraged to look for different tessellations for each of these shapes. They should find that with these shapes, as with those above, the sum of the angle measures around the point is always 360°.

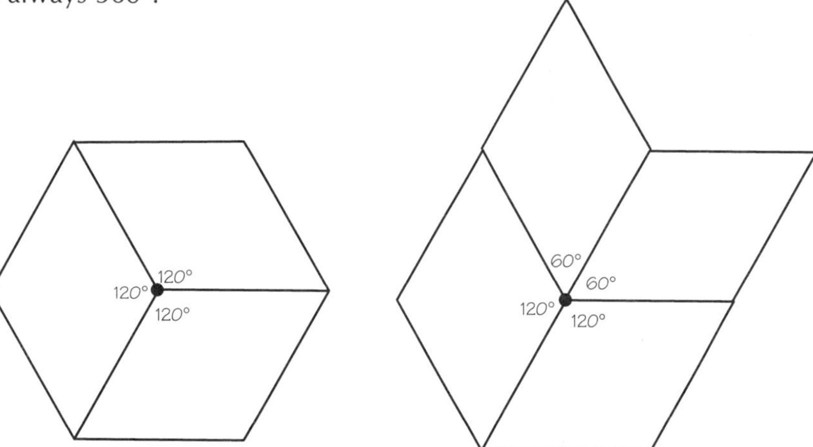

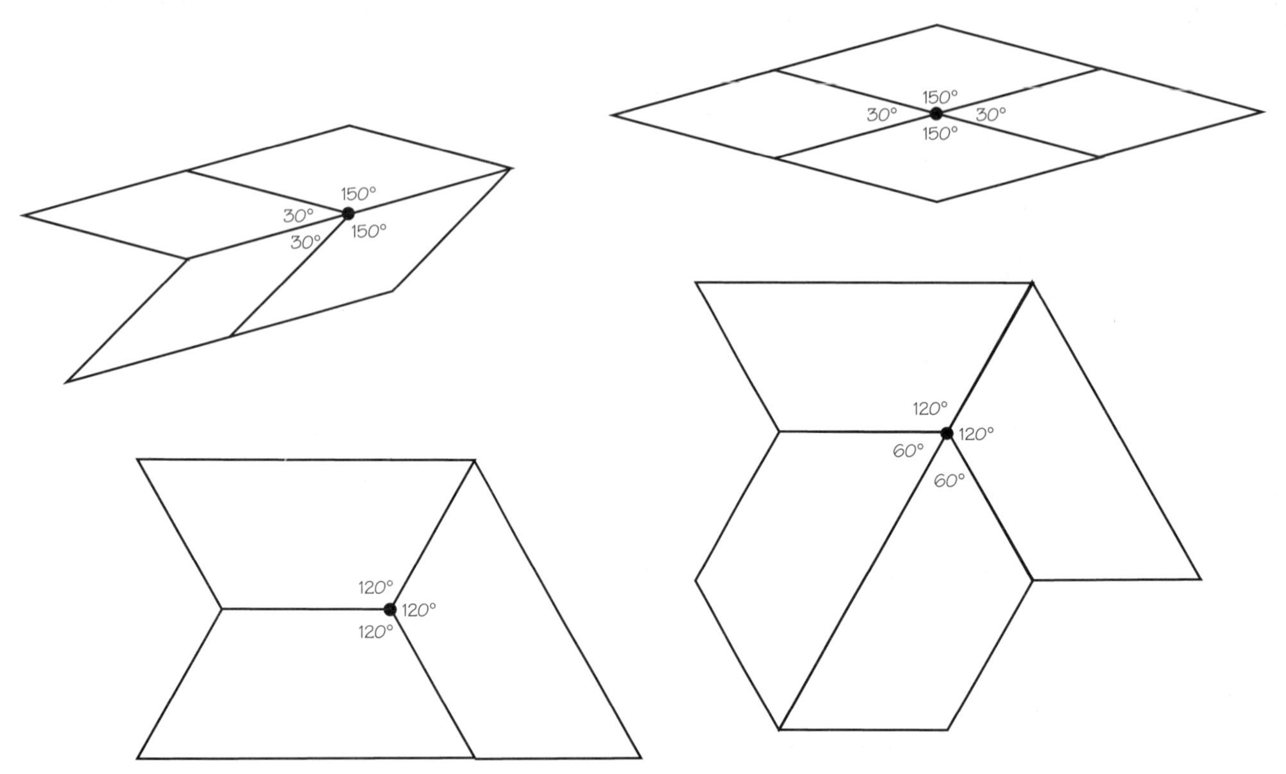

In Part 2, as they choose combinations of Pattern Blocks to form their tessellations, students may find it helpful to use the fact that the polygons needed to completely surround a point must have angle measures totaling 360°. For example, they may start with a triangle and a hexagon (for a total of 180° at the point), and then see which other Pattern Block shapes could be used to fill out the remaining 180°.

Once they choose the shapes, they must then select a way of arranging them. Students may discover that several different designs can be formed using the same pieces. As they extend their designs, students may find that not all tessellations will form mathematical mosaics. Students will need to check to see whether the arrangement of the shapes is exactly the same at each vertex in the design. Several examples (some mathematical mosaics, some not) are shown below.

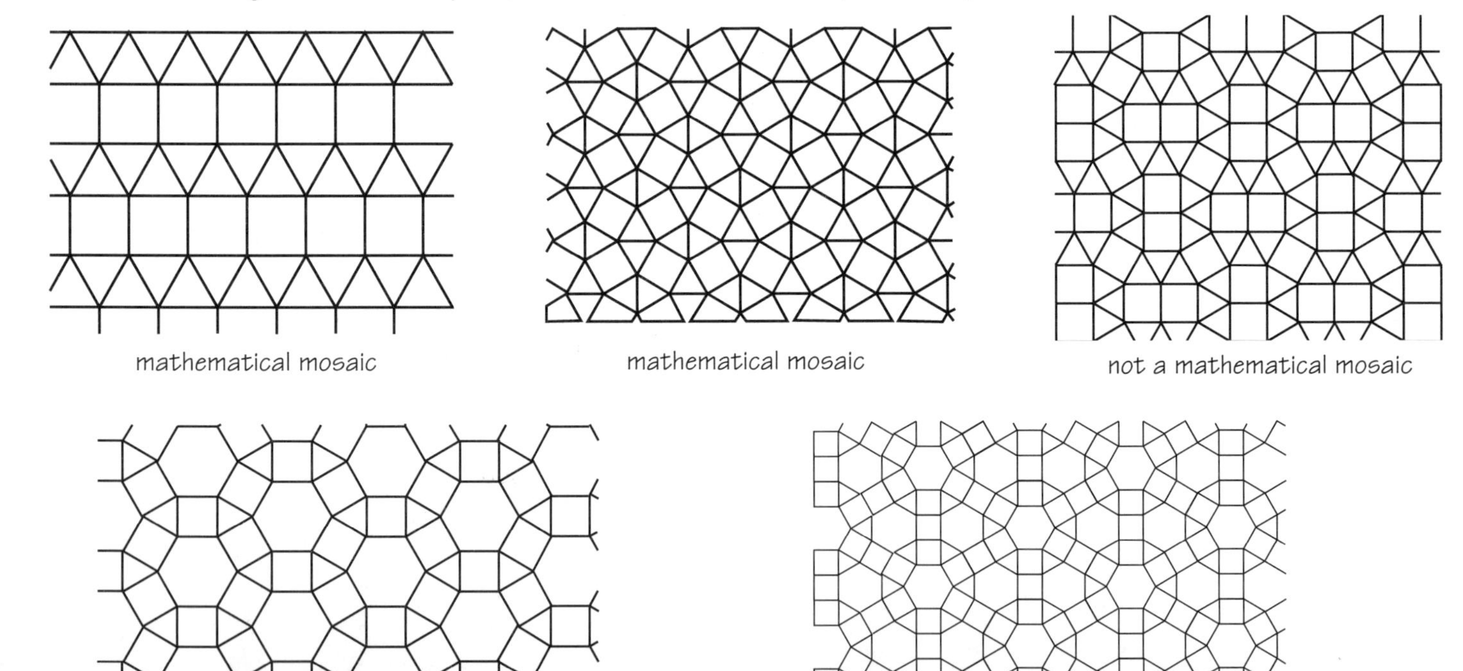

Investigating Solid Geometry

The four activities in this cluster use Snap Cubes as a forum for learning about geometric solids. The first two lessons give students practice in creating two-dimensional representations of three-dimensional figures; they also examine various rotations and reflections of those figures. The second two lessons focus on physical attributes of polyhedra, such as numbers of vertices, edges, and faces, and the functional relationships among those numbers.

1. Modular Seating Cubes (*Exploring representations of three-dimensional shapes*)

In this activity students use isometric dot paper to create several different views of a single three-dimensional shape.

The *On Their Own* instructions do not include directions for drawing three dimensions on (two-dimensional) paper. This leaves it to the teacher to decide whether to introduce the activity with a practice drawing at the board (or at students' desks), or to have students discover a method on their own. Teachers who do want to introduce the activity might start with simple cubes or rows of cubes (as shown here), and have students figure out multidirectional shape representations on their own.

2. Pentacube and Hexacube Twins (*Extending representations of three-dimensional shapes*)

In this activity students create five-cube and six-cube structures and examine their reflections. In drawing representations of these structures on isometric dot paper, they reinforce skills introduced in the previous activity.

On Their Own Part 1 assumes an understanding of mirror images or reflections.

Where's the Mathematics? provides hints for drawing reflections of three-dimensional representations.

3. Slice 'n' Dice Cubes (*Exploring surface and volume relationships in rectangular solids*)

In this activity, students explore number relationships among the vertices, edges, faces, and volume of cubes and rectangular prisms, with the emphasis on spatial visualization. The activity provides a good opportunity for organizing and analyzing data, and for making predictions and creating algebraic formulas based on data analysis.

On Their Own Part 2 assumes that students are familiar with the term "rectangular prism."

4. Saving Paper (*Exploring geometric solids*)

In this activity, students examine solids made from cubes, charting numbers of vertices, edges, and faces in order to derive formulas based on their observations. They build nets for cubes and determine the efficiency of the various nets.

Where's the Mathematics? (page 86) describes Euler's formula relating numbers of vertices, edges, and faces in a polyhedral solid.

MODULAR SEATING CUBES

- Three-dimensional shapes
- Perspective
- Visual perception

Getting Ready

What You'll Need

Snap Cubes, 5 per student, each a different color

Isometric dot paper, page 111

Colored pencils or crayons in colors of cubes

Activity Master, page 102

Overview

Students build Snap Cubes structures, study them from various angles, and then record what they see using isometric dot paper. In this activity, students have the opportunity to:

- focus on visual perspective
- represent three-dimensional objects in a two-dimensional plane
- use isometric dot paper as a tool for making perspective drawings

Other *Super Source* activities that explore these and related concepts are:

Pentacube and Hexacube Twins, page 74

Slice 'n' Dice Cubes, page 78

Saving Paper, page 83

The Activity

On Their Own (Part 1)

Elena is rearranging the furniture in her room. In addition to her bed and a dresser, she has 4 different-colored modular seating cubes. She would like to consider possible arrangements of the seating cubes without actually moving them. Can you help Elena by drawing different views of possible arrangements?

- Use 4 different-colored Snap Cubes to represent the 4 modular seating cubes. Connect the 4 Snap Cubes in any way you choose to model a potential seating arrangement for Elena's room.
- Using isometric dot paper, draw as many different views of your 4-cube arrangement as possible. Color your drawings, showing the color of each cube in your model.
- Compare your cube arrangement and isometric drawings with those of other members of your group.
- Be ready to explain how your drawings show three dimensions.

Thinking and Sharing

Have volunteers share their models showing potential arrangements of the four seating cubes. Taking one arrangement at a time, ask students with the same arrangement to share and compare their isometric drawings. Have students hold up their cube structures in the positions that match each two-dimensional drawing.

Use prompts like these to promote class discussion:

- How did you decide on your arrangement of the seating cubes?
- How many different dot-paper drawings of your 4-cube arrangement do you think are possible? Explain.
- What was difficult about the activity? What was easy? Why?
- What views of the cube arrangement could not be drawn on isometric dot paper? Why couldn't they be drawn?

On Their Own (Part 2)

What if... ***Elena bought a fifth seating cube for her room? Could you design a 5-cube seating arrangement and draw different views of it so that Elena could arrange the seating cubes working from your drawings?***

- Use 5 different-colored Snap Cubes to represent the 5 modular seating cubes. Connect the 5 Snap Cubes in any way you choose to model Elena's new seating arrangement. Keep your model hidden from your partner.
- Using isometric dot paper, draw at least 4 different views of your 5-cube arrangement. Color your drawings, showing the color of each cube in your model.
- Exchange your dot-paper drawings with a partner. Using the drawings, try to build each other's seating arrangements. Then draw a view of the arrangement that is different from the views your partner drew.

Thinking and Sharing

Invite students to show their models and to describe how they interpreted their partner's dot-paper drawings to build the seating arrangement.

Use prompts like these to promote class discussion:

- How did you decide on the arrangement of your seating cubes?
- What was difficult about making the dot-paper drawings? What was easy? Why?
- How did you go about building your partner's seating arrangement with the Snap Cubes?
- Was the information supplied in the dot-paper drawings sufficient to build the Snap Cube model? Explain why or why not.
- After building your partner's model, how did you find a new view to draw?

Write about the methods you used to draw your models on dot paper. Describe any differences that exist between the actual models and the two-dimensional representations, such as relative side lengths, angle measures, and changes in shape. You may want to use diagrams to illustrate your explanations.

Teacher Talk

Where's the Mathematics?

It is important for students to be able to draw representations of three-dimensional objects in order to be able to communicate about them. By using isometric dot paper, students can more readily record their views in two dimensions. Shading or coloring can further enhance the drawings and make it easier to distinguish particular faces of the cubes in the structure.

Not all views of a cube arrangement can be represented on isometric dot paper. For example, a 4-cube L-shaped structure, when viewed "straight on" would look like this:

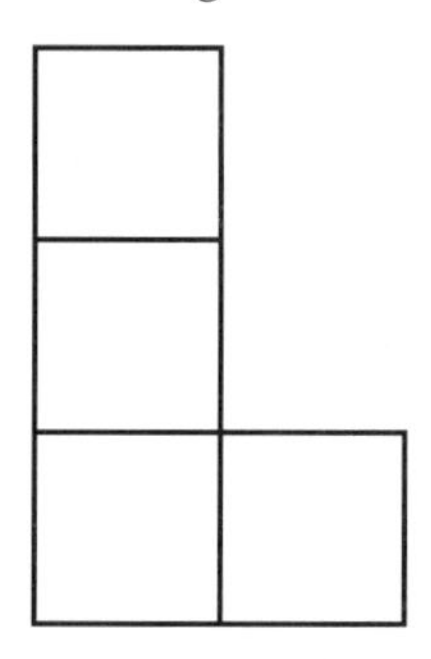

From different sides, the structure would offer one of these views:

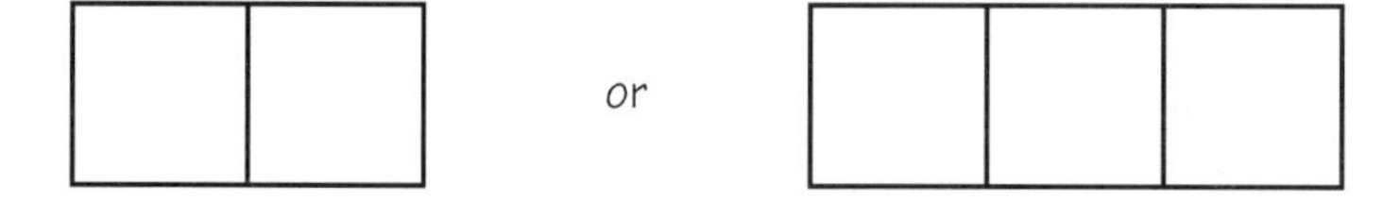

Note that none of the three views pictured above can be drawn on isometric dot paper.

As they work with the isometric dot paper, students will discover the kind of views of their models that *can* be drawn using the dot paper. They may also find it easier to draw their representations if they hold their Snap Cube models in positions where the vertices of the cubes are oriented in the same way as the dots on the dot paper. The edges of their cubes will then correspond to line segments that can be drawn by connecting the dots on the paper.

Some students may have difficulty drawing their Snap Cube arrangements. For these students, it may be necessary to start by making dot-paper drawings of models made up of one, two, or three cubes attached end-to-end in a single row. As their drawing skills improve, it will become easier for them to increase the number of cubes and the complexity of the arrangements.

Students may find some views easier to draw than others. The drawings below show several views of both 4- and 5-cube models. Drawings marked with the same letter show different views of the same arrangement.

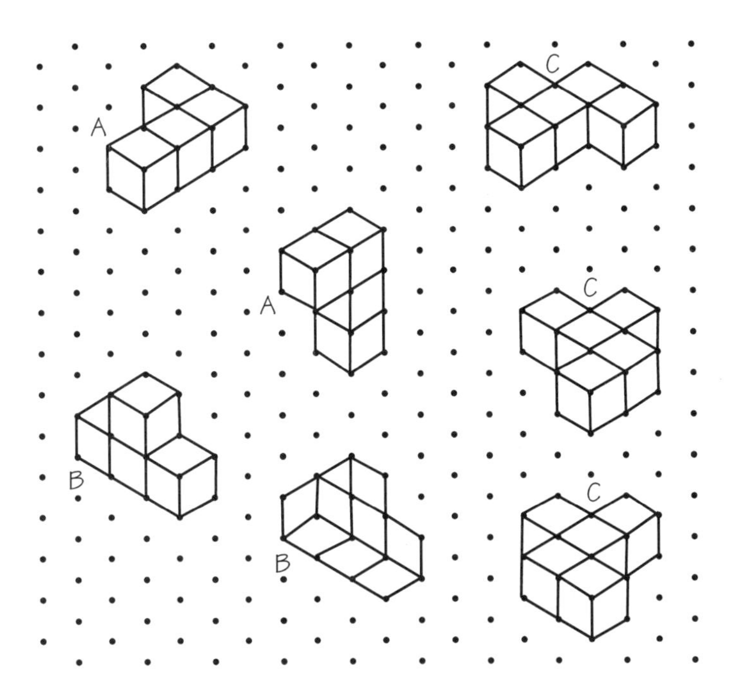

As they compare their drawings to their models, students may notice that certain relationships on the original model, such as parallelism of edges and parallelism of faces, remain unchanged when viewed on the dot-paper drawings. However, the shapes of the faces, when drawn, are not squares. Their recognition of the differences between the actual model and its representation may help students understand how their drawings provide perspective and give the illusion of three dimensions.

PENTACUBE AND HEXACUBE TWINS

- Spatial visualization
- Transformational geometry
- Congruence

Getting Ready

What You'll Need

Snap Cubes, 100 per pair

Isometric dot paper, page 111

Mirrors (optional)

Activity Master, page 103

Overview

Students use Snap Cubes to build 5-cube and 6-cube structures which, no matter how they are positioned when placed on a table, have at least one cube that does not touch the table. They then identify which of their structures are reflections of each other and draw them on isometric dot paper. In this activity, students have the opportunity to:

- strengthen visual perception
- represent three-dimensional objects in two dimensions
- draw and assemble models that are reflections of given structures
- differentiate between reproductions and reflections
- use isometric dot paper as a tool for verifying reflections of three-dimensional structures

Other *Super Source* activities that explore these and related concepts are:

Modular Seating Cubes, page 70

Slice 'n' Dice Cubes, page 78

Saving Paper, page 83

The Activity

On Their Own (Part 1)

A pentacube is a Snap Cube structure made from 5 cubes. A pop-up pentacube is a pentacube which, no matter how it is placed on a table, will have at least one cube that does not touch the table. How many different pop-up pentacubes can you make?

- Working with a partner, build as many pop-up pentacubes as you can. Check to make sure that each of your structures, no matter how it is positioned on your table, has at least one cube that does not touch the table.
- Compare your pentacubes and eliminate any duplicates. Rotating, flipping, and turning the structures may help you identify pentacubes that are identical.
- Pop-up pentacube twins are pentacubes that are mirror images of each other. Look at your models and see if any of them are pop-up pentacube twins. Draw each set of pentacube twins on isometric dot paper, drawing them in positions that make it easy to see that they are reflections of each other.
- Draw your other pop-up pentacubes on a separate piece of isometric dot paper. Try to figure out why these pentacubes have no twins.

Thinking and Sharing

Invite pairs of students to present one or two of their pairs of pentacube twins until all six pairs have been presented. Then ask students to share any "twinless" pentacubes that they found. Have students study the structures on display and identify duplicates, missing models, or structures that do not satisfy the requirements of a pop-up pentacube.

Use prompts like these to promote class discussion:

- How did you go about creating your pop-up pentacubes?
- Did you use a strategy to find new ones? If so, explain what you did.
- Did looking for pentacube twins help you find others that were missing? If so, explain.
- How many pop-up pentacubes did you find? Do you think you found them all? Why or why not?
- How did you discover pentacubes that were duplicates of each other?
- What was difficult about making the dot paper drawings? What was easy? Why?
- What strategies did you use to draw pentacubes that were reflections of each other?
- What did you notice about pentacubes that do not have twins?

On Their Own (Part 2)

What if... ***you had a drawing of a pop-up hexacube? Could you draw and build its twin?***

- Using 6 Snap Cubes, build a pop-up hexacube. Keep your hexacube hidden from your partner.
- Draw your pop-up hexacube on isometric dot paper. Exchange drawings with your partner.
- Working from your partner's drawing, draw the mirror image (reflection) of the structure on the isometric dot paper.
- Now build the two hexacube structures (the original and its reflection). Examine them from all perspectives to make sure that they are hexacube twins. Then check your hexacubes with the original structure made by your partner.

Thinking and Sharing

Have students share their pop-up hexacube twins and drawings. Have them explain how they know their hexacubes are twins.

Use prompts like these to promote class discussion:

- What methods did you use to draw the reflection of your partner's drawing?
- What was difficult about making the dot paper drawing of the reflection? What was easy? Why?
- How did you go about building the hexacube twins from the drawings?
- What was difficult about building them? What was easy? Why?
- How did your hexacubes compare to your partner's original structure?

Write about the differences and similarities between a reflection and a reproduction. You may find it helpful to use a photograph of yourself and a mirror to make the comparisons. Include any diagrams that might help illustrate your points.

Teacher Talk

Where's the Mathematics?

As students build and collect their pop-up pentacubes, they may begin to realize the challenge involved in searching for duplications and reflections. Students will need to consider different perspectives of their structures as they flip, turn, and rotate them for comparison.

The 17 possible pop-up pentacubes are shown below. There are six pairs of pentacube twins: A and B, C and D, E and F, G and H, I and J, and K and L. Each of these twins is the reflection (mirror image) of the other, but is not identical, or congruent, to the other. Five of the pentacubes appear to have no twins: M, N, O, P, and Q. These structures can be rotated to produce their own reflections, therefore each is congruent to its mirror image.

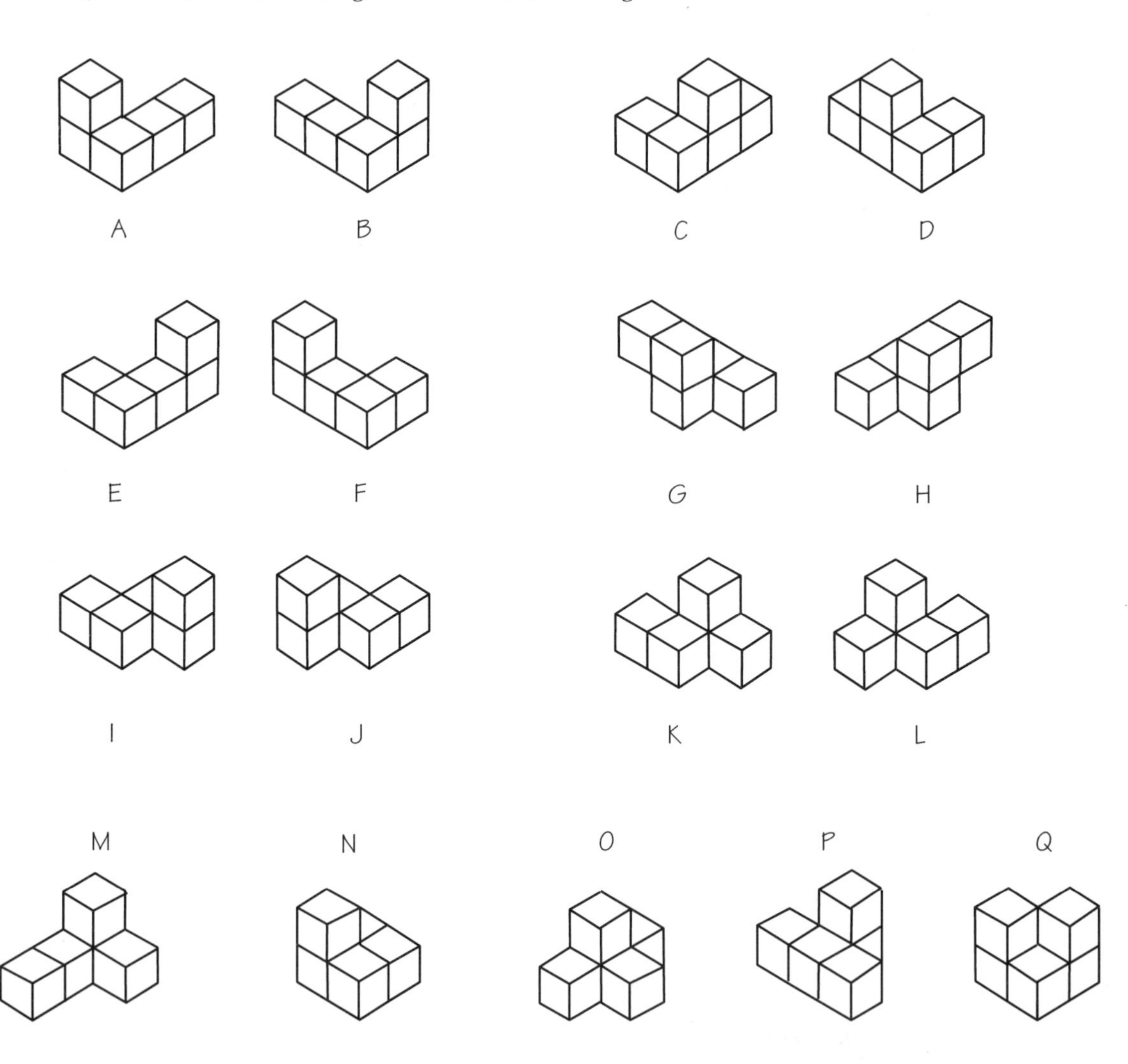

Some students may start building pentacubes in a random manner. Other students may build one pentacube and then move one of its cubes to a new position to form a second pentacube. Using this new structure, they may again move one cube to generate a third pentacube, and so on. Some students may choose to build a pentacube and then work on building its reflection (mirror image), generating a pair of pentacube twins each time. Positioning a small mirror to the side of a pentacube to show its mirror image might help students who have difficulty visualizing the reflection.

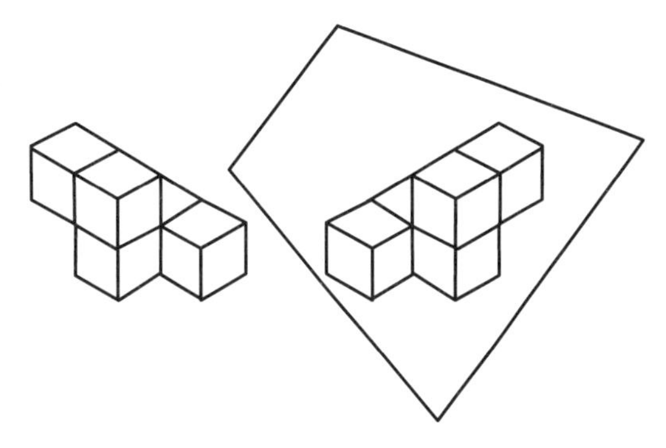

To draw their pentacubes and hexacubes on isometric dot paper, students may find it helpful to position their structures so that the corners of the cubes correspond to the dots on the paper. The lines that they draw connecting these dots will then represent the edges of the Snap Cubes. Once a drawing has been made, students may choose to use a small mirror to see its mirror image before attempting to draw its reflection.

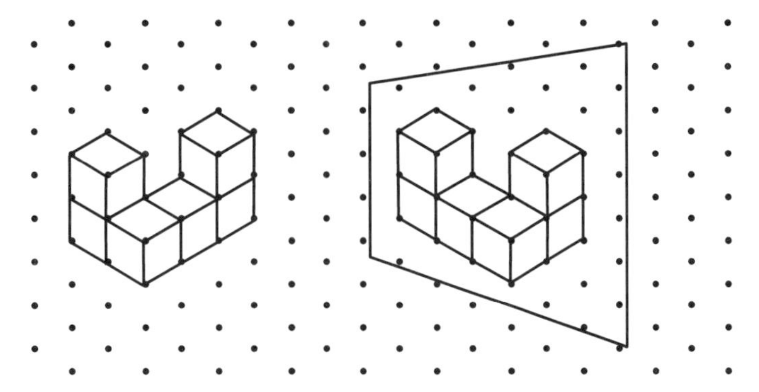

Some students may discover that the reflection of a drawing can be found by folding the dot paper. The paper must be folded so that the dots align; then the drawing can be traced first on the back of the paper (with the paper folded) and then onto the front (retracing the new drawing from the back).

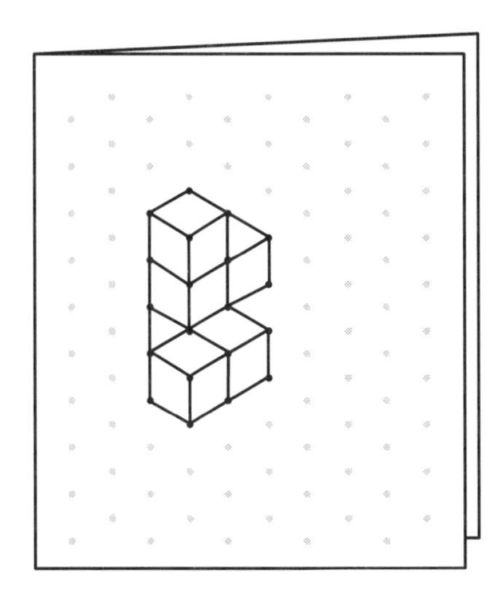

SLICE 'N' DICE CUBES

- *Surface area*
- *Volume*
- *Organizing data*
- *Algebra*

Getting Ready

What You'll Need

Snap Cubes, about 150 per pair

Stick-on dots (optional)

Activity Master, page 104

Overview

Students form cube and rectangular prism structures using Snap Cubes and imagine them being dipped in paint. They then investigate the patterns formed by the numbers of individual cubes that have paint on a given number of faces. In this activity, students have the opportunity to:

- investigate the relationship between volume and surface area of cubes and rectangular prisms
- assess the number of vertices, edges, and faces of cubes and rectangular prisms
- collect and organize data
- analyze number patterns and use them to make predictions
- write algebraic formulas

Other *Super Source* activities that explore these and related concepts are:

Modular Seating Cubes, page 70

Pentacube and Hexacube Twins, page 74

Saving Paper, page 83

The Activity

On Their Own (Part 1)

Laura constructed some cube structures using Snap Cubes. These structures, which she called Slice 'n' Dice Cubes, measured 2x2x2, 3x3x3, and 4x4x4. Once the Slice 'n' Dice Cubes were assembled, they were dipped in red paint and allowed to dry. When Laura took these cubes apart, she was surprised to find that there were relationships between the dimensions of the cube structures and the number of cubes with paint on exactly 0, 1, 2, or 3 faces. What did she discover?

- Working with a partner, build your own Slice 'n' Dice Cubes measuring 2x2x2, 3x3x3, and 4x4x4. Record the number of Snap Cubes used to build each cube structure, and the number of its vertices (corners), edges, and faces.
- Imagine dipping your Slice 'n' Dice cubes in red paint. (Optional: Use stick-on dots to mark the painted faces of each Snap Cube.)

- For each structure, determine the number of individual Snap Cubes that have paint on exactly 3 faces, on exactly 2 faces, and on exactly 1 face. Determine the number of cubes with no painted faces.
- Organize and record your data. Look for patterns.
- Take apart the three cube structures. Build a new structure whose dimensions are 5x5x5. Imagine dipping it in red paint. Investigate this structure as you did the others and record your data.
- Look to see if you can identify any relationships between the dimensions, the total number of cubes, the number of different types of painted cubes, and the number of unpainted cubes. Be ready to explain your findings.

Thinking and Sharing

Invite volunteers to tell about how they investigated their Slice 'n' Dice Cubes and to describe any patterns they found. Then create a class chart with the headings *Dimensions of Cube, # of Snap Cubes, # of vertices, # of edges, # of faces, # of cubes with exactly 3 painted faces, # of cubes with exactly 2 painted faces, # of cubes with exactly 1 painted face, # of cubes with no painted faces.*

Use prompts like these to promote class discussion:

- Where were the cubes with exactly 3 painted faces located? Where were the cubes with exactly 2 painted faces located? (etc.)
- How did you organize your data?
- What data were the same for each cube?
- What patterns did you find?
- What patterns do you see in the class chart?
- How are the dimensions of the cube related to the number of cubes that have paint on 3 (2, 1, 0) faces?
- What predictions could you make about a Slice 'n' Dice Cube measuring 10 x 10 x 10?
- What predictions could you make about a Slice 'n' Dice Cube that has *n* Snap Cubes per edge?

On Their Own (Part 2)

What if... ***Laura decided to build different-shaped rectangular prisms using Snap Cubes? Would she find the same relationships as those she discovered for the Slice 'n' Dice Cubes?***

- Working with your partner, build at least 3 rectangular prisms using a different number of Snap Cubes for each one. Record the dimensions of each prism and the number of its vertices, edges, and faces.
- Imagine each structure being dipped in red paint. (Optional: Use stick-on dots to mark the painted faces of each Snap Cube.)

- For each prism, determine the number of individual Snap Cubes that have paint on exactly 3 faces, on exactly 2 faces, and on exactly 1 face. Determine the number of cubes with no painted faces.
- Organize and record your data. Look for patterns.
- Look to see if you can find any relationships between the dimensions of the prisms and the number of cubes with paint on a given number of faces. How do these relationships compare to those you found for the Slice 'n' Dice Cubes? Be ready to explain your findings.

Thinking and Sharing

Have students share the prisms they made. Construct a chart similar to the one used in the first part of the activity to organize the data. Column headings *top and bottom faces, left and right faces,* and *front and back faces* may be added to the chart to identify where specific painted cubes were found.

Use prompts like these to promote class discussion:

- What did you discover about the faces of the prisms?
- Where were the cubes with exactly 3 painted faces located? Where were the cubes with exactly 2 painted faces located? (etc.)
- How did you organize your data?
- What data were the same for each prism?
- What patterns did you find?
- What patterns do you see in the class chart?
- How are the dimensions of the prism related to the number of cubes that have paint on 3 (2, 1, 0) faces?
- What predictions could you make about a prism with dimensions 3 x 5 x 8?
- What predictions could you make about a prism that has a length of l units, a width of w units, and a height of h units?

Write a letter to Laura explaining how to figure out the number of Snap Cubes with paint on at least 2 painted faces in a rectangular prism whose dimensions are 4x6x11.

Teacher Talk

Where's the Mathematics?

Regardless of the dimensions of a Slice 'n' Dice Cube, there are 8 vertices, 12 edges, and 6 faces on the structure. The cubes with exactly 3 painted faces are located at the vertices; the cubes with exactly 2 painted faces are located on the edges between the vertices; the cubes with exactly 1 painted face are located on the faces but not on the edges or at the vertices; and the cubes with no painted faces are located at the core of the cube, beneath all painted cubes. (Note: It is important for students to understand the meaning of the phrase *exactly two painted faces* as compared to *two painted faces*, as cubes with 3 painted faces also satisfy the condition of having 2 painted faces.)

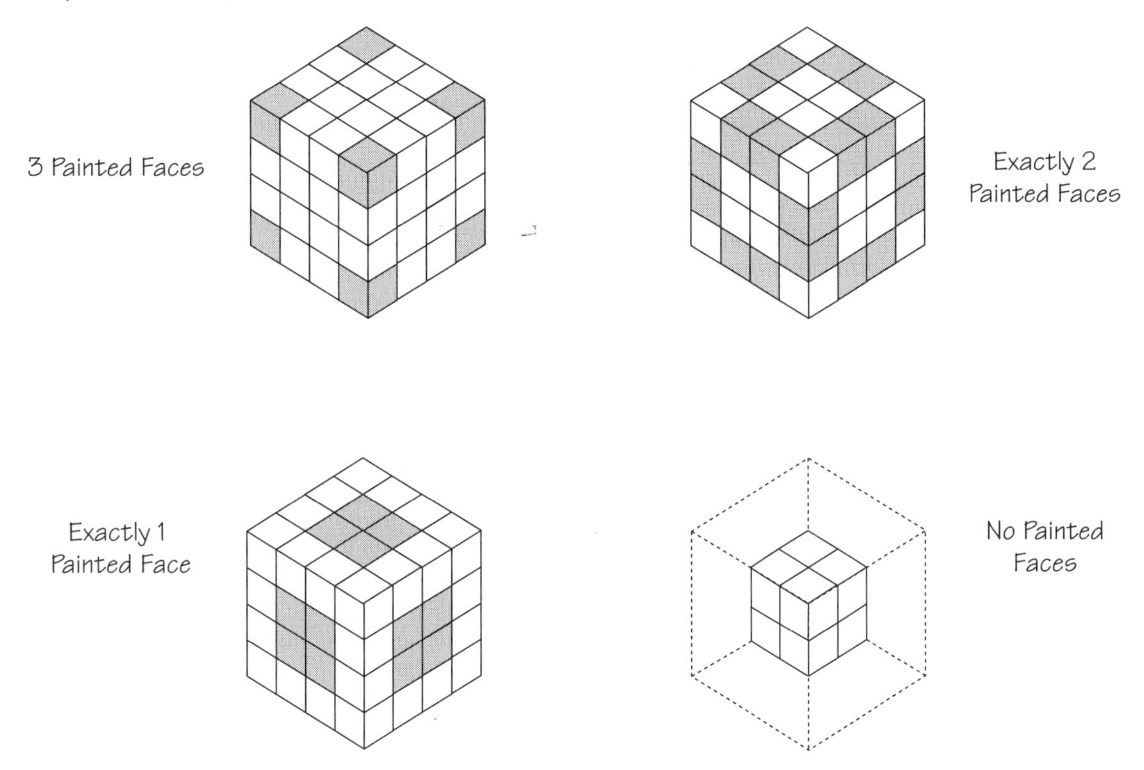

The data reveal a number of interesting patterns and relationships.

Dimensions of Cube	# of Snap Cubes	# of cubes with exactly 3 painted faces	# of cubes with exactly 2 painted faces	# of cubes with exactly 1 painted face	# of cubes with no painted faces
2 x 2 x 2	8	8	0	0	0
3 x 3 x 3	27	8	12	6	1
4 x 4 x 4	64	8	24	24	8
5 x 5 x 5	125	8	36	54	27

Students should recognize that the total number of Snap Cubes used to build each structure is the product of its dimensions. This product represents the volume of the cube: Volume = $n \times n \times n$, or n^3, where n represents the number of Snap Cubes per edge. Knowing the volume of each Slice 'n' Dice Cube can be useful in helping students check their work, as the number of painted and unpainted cubes must equal the total number of cubes in the structure.

Students should also recognize that for any size cube, there will always be 8 Snap Cubes with paint on exactly 3 faces, namely, the cubes that are located at the vertices of the cube structure. The Snap Cubes with exactly 2 painted faces are located on the edges between the vertices. In a 5 x 5 x 5 Slice 'n' Dice Cube, there are 3 such cubes per edge. Since there are 12 edges on each Slice 'n' Dice Cube, the total number of cubes with paint on exactly 2 faces is 12 x 3, or 36. In an $n \times n \times n$ Slice 'n' Dice Cube, there are $(n - 2)$ such cubes per edge, and, therefore, a total of $12 \times (n - 2)$ cubes with exactly 2 painted faces on the entire cube structure.

The Snap Cubes with exactly 1 painted face are located on each face but not on the edges or at the vertices. In a 5 x 5 x 5 Slice 'n' Dice Cube, there are 3 x 3, or 3^2, such cubes per face. Since there are 6 faces on each Slice 'n' Dice Cube, the total number of cubes with exactly 1 painted face is 6×3^2, or 54. In an $n \times n \times n$ Slice 'n' Dice Cube, there are $(n - 2) \times (n - 2)$, or $(n - 2)^2$, such cubes per face, and, therefore, a total of $6 \times (n - 2)^2$ cubes with exactly 1 painted face on the entire cube structure.

The Snap Cubes with no painted faces are located under the painted shell, at the core of the Slice 'n' Dice Cube. When the painted Snap Cubes are "peeled" off the Slice 'n' Dice Cube, each dimension is decreased by 2 units. For example, if the outer shell of a 5 x 5 x 5 cube were peeled off, a 3 x 3 x 3 core (consisting of unpainted Snap Cubes) would be revealed. The number of Snap Cubes in this inner cube would be 3^3, or 27. If the painted Snap Cube shell is peeled off an $n \times n \times n$ Slice 'n' Dice Cube, a core whose dimensions are $(n - 2) \times (n - 2) \times (n - 2)$ would be revealed. The number of Snap Cubes in this inner cube would be $(n - 2) \times (n - 2) \times (n - 2)$, or $(n - 2)^3$.

Students may recognize that they can assemble these algebraic expressions into a statement or equation that relates the numbers of painted and unpainted Snap Cubes to the volume of the cube structure.

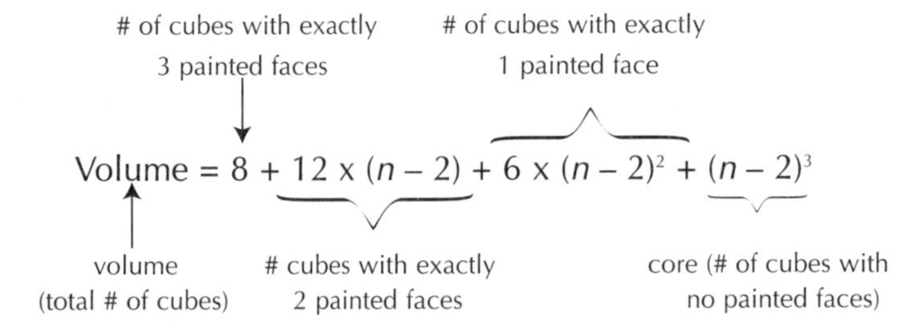

Unlike a cube, a rectangular prism may have three pairs of faces with different dimensions. Students should find that there are, again, 8 Snap Cubes with paint on exactly 3 faces (the cubes at the vertices). The cubes with paint on exactly 2 faces, 1 face, or no faces will be located in the same positions as on the cube structure. However, these numbers will vary from edge to edge and face to face, as the length of the edges and the dimensions of the faces will differ. For example, in a rectangular prism with dimensions 3 x 5 x 8, there are four edges of length 3, four edges of length 5, and four edges of length 8. There are two faces with dimensions 3 x 5, two faces with dimensions 5 x 8, and two faces with dimensions 3 x 8. The core prism has dimensions 1 x 3 x 6. Students can use their recognition of the significance of the differing dimensions to determine the numbers of cubes with a given number of painted faces in a given prism, and to modify the above formula to reflect the variable measures of length, width, and height.

SAVING PAPER

- Nets
- Spatial visualization
- Surface area
- Euler's formula

Getting Ready

What You'll Need

Snap Cubes, 8 per pair

Snap Cube grid paper, page 112

Scissors

Tape

Activity Master, page 105

Overview

Students build 4-cube structures made from Snap Cubes, and design nets that could be used to cover the structures. They study and compare their nets, and search for efficient ways to cut multiple copies of them from cardboard rectangles. In this activity, students have the opportunity to:

- strengthen spatial visualization in two and three dimensions
- enhance their understanding of surface area
- design and assemble nets
- discover Euler's Formula

Other *Super Source* activities that explore these and related concepts are:

Modular Seating Cubes, page 70
Pentacube and Hexacube Twins, page 74
Slice 'n' Dice Cubes, page 78

The Activity

On Their Own (Part 1)

Juan works in the shipping department of a company that manufactures coffee mugs, each of which is packaged in its own cubic cardboard box. Juan is filling an order for 4 coffee mugs. He needs to determine how to group the 4 boxes together as a single unit for shipping and cut one piece of wrapping paper that can be used to cover the 4 boxes. How might he do this?

- Use 4 Snap Cubes to represent the 4 boxes containing the coffee mugs. Working with your partner, decide on 2 different ways to group the boxes together for shipping.
- Using Snap Cube grid paper, design 2 nets for each of your structures. Think of your nets as the wrapping paper that could be used to cover the structures without any overlaps.
- Cut out each net to see if it works. Fold on the grid paper lines and tape the common edges together. If the net doesn't work, revise the design until it does.
- Determine the number of vertices, edges, and faces on each of your "wrapped packages." Record your results.
- Make a clean copy of each net to share with the class. Be ready to show how you figured out the design for your nets.

Thinking and Sharing

Invite pairs to share their Snap Cube arrangements and to post their corresponding nets. For each of the structures displayed, ask whether other pairs designed a different net, and, if so, add them to the display. Make a class chart with the headings: *Number of Vertices (V), Number of Edges (E)*, and *Number of Faces (F)*. Have students record their data for their different structures in the class chart.

Use prompts like these to promote class discussion:

- How did you decide on your Snap Cube arrangements?
- How did you go about designing the nets for your structures? What was the hardest part? What was the easiest part?
- Did you have to design more than one net before you found one that worked? If so, what kinds of changes did you have to make?
- How many squares of grid paper were required to make each of your nets?
- How many vertices, edges, and faces were on your "wrapped packages"? How did you go about finding these numbers?
- Do you see any relationships among the number of vertices, edges, and faces in each structure? If so, describe them.
- Look at the different nets for your two structures. Imagine tracing them onto a rectangular piece of grid paper. Which would require the smallest rectangle? What would the dimensions of the rectangle be?

On Their Own (Part 2)

What if... *Juan has to fill 4 of these orders for sets of 4 coffee mugs? How might he arrange 4 copies of the net on a rectangular sheet of wrapping paper if he wants to minimize the amount of wasted paper?*

- Choose one of your 4-cube structures to represent your package. Compare all of the nets for the structure. Choose one of the nets (different from the one chosen by your partner) and make 4 copies of it.
- Using one sheet of Snap Cube grid paper, find a way to enclose the four nets in a rectangular area. Try to arrange your nets in such a way as to minimize the amount of wasted paper.
- Find the area of the rectangle enclosing the 4 nets and the area of the wasted paper.
- Compare your work with your partner's to determine which net and which arrangement saves the most paper. If you think a different net or arrangement might be more efficient, try it and see.

Thinking and Sharing

Have students share their work, posting their grid-paper drawings and explaining their conclusions.

Use prompts like these to promote class discussion:

- How did you choose the net you thought best to use?
- How did you go about arranging the 4 nets on the rectangular grid paper?
- Did you find that flipping, rotating, or turning the nets was helpful in arranging them on the paper? Explain.
- How did the area of the rectangle you used compare to the area of the rectangle used by your partner?
- How did the areas of the wasted paper compare? Which net made for the least amount of waste? Explain why.
- Did you experiment with a third net? If so, tell about your findings.
- Look at the class results. What observations and conclusions can you make?

Write about the methods you would use to construct a net for any 3-dimensional structure. Include any diagrams that might help illustrate your methods.

Teacher Talk

Where's the Mathematics?

There are seven possible structures that can be built using four Snap Cubes. Students' models of 4-mug packages should match, or be a transformation of, one of the structures pictured here:

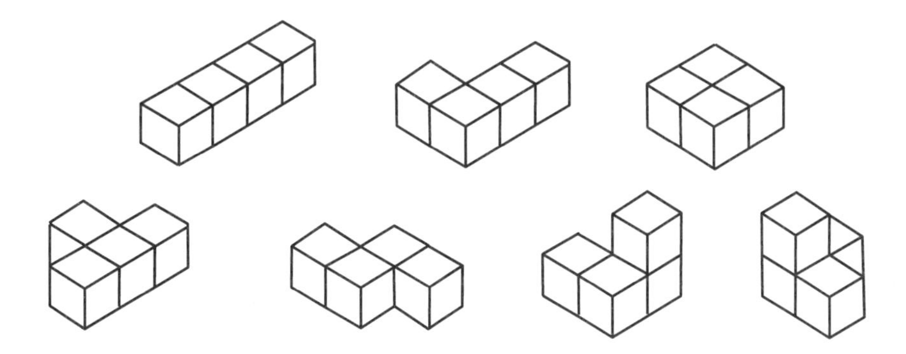

Creating nets for their packages will require students to use spatial visualization skills. Some students may find it beneficial to think about "unfolding" their structure to two dimensions, while others may focus more on the idea of "wrapping" the structure. Many students will need to experiment by making a net, assembling it, and comparing it to their original structure. Once the net is assembled, if students find that the net does not match the original 4-cube structure, they may be able to cut off misplaced squares and attach them to the remaining net along the appropriate edges, taping them in place. The adjusted net can then be unfolded and redrawn on grid paper.

Students may find that the hardest part of designing the net is visualizing where the squares used to cover the top of the 4-cube structure should be located in the net. Squares for covering the bottom of the structure can be found by laying the 4-cube structure on the paper and tracing around the shape. Additional squares for the net can then be determined by lifting the paper around the shape to find those squares that will cover the sides. This method may also help students to see where to attach the squares for the top.

Students may be surprised at the number of different nets generated by their classmates for each 4-cube arrangement. Some of the nets may vary greatly, while others may vary by only one or two squares, such as in the example below, where the square marked with an "x" is the only difference between the two nets.

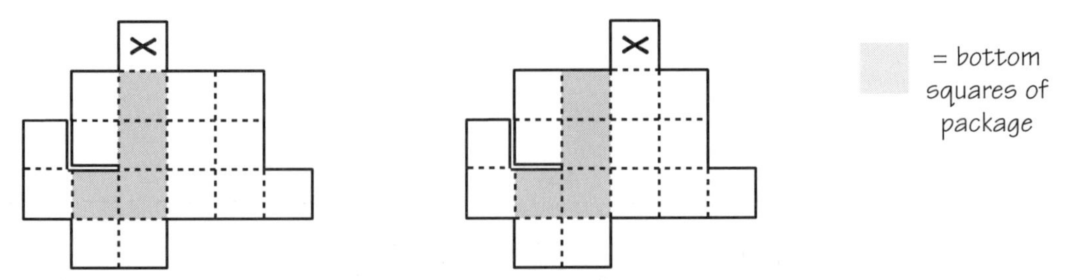

Students should find that the number of vertices *(V)*, edges *(E)*, and faces *(F)* of their structures will vary, depending upon the arrangement of the 4 cubes. For example, the structure in figure A below has 8 faces, 12 vertices, and 18 edges, while the structure in figure B has 10 faces, 16 vertices, and 24 edges. By examining the data in the class chart, students may discover a relationship that is satisfied by each of these sets of values: $F + V = E + 2$. This relationship is known as *Euler's Formula.*

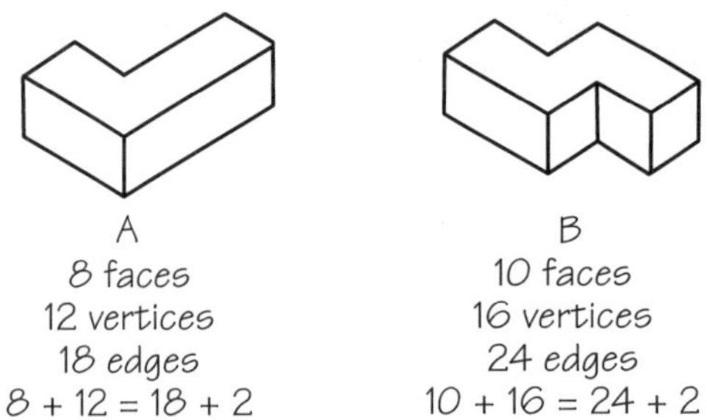

Students may be interested in investigating whether Euler's Formula holds true for any 3-dimensional shape.

In Part 2, ways to minimize the area of the rectangle needed for the 4 nets and the amount of wasted paper will vary depending upon the structure. In general, nets that can be arranged so that they share many common edges will produce the least amount of waste. One such arrangement is shown below.

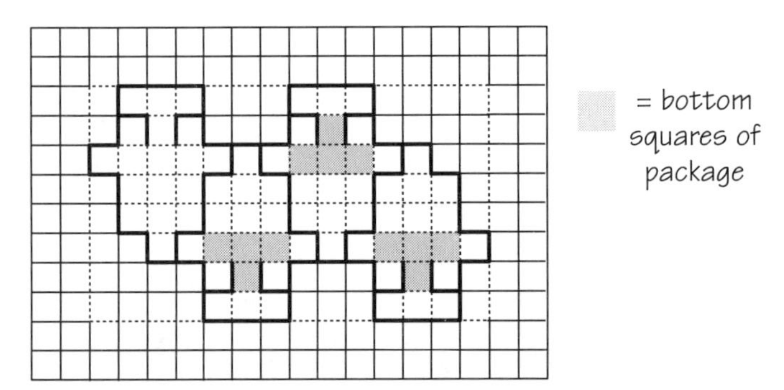

As students compare their results with their classmates', ideas for creating even more efficient arrangements may come to light. Students may be interested in experimenting with this concept further, working with different structures and their various nets.

Blackline Masters

Cardboard Cartons

Part 1

Tino has a large sheet of cardboard and 6 square tiles that measure 12 inches on each edge. How many different 6-tile arrangements (hexominoes) can Tino trace on the cardboard sheet if at least one complete edge of each tile must touch one complete edge of another tile?

- Working with a partner, use 6 Color Tiles to make as many hexominoes as you can.
- Record your models on 1-inch grid paper, cut them out, and decide on a way to sort them.
- Make sure that each hexomino is different from the others. Eliminate hexominoes that are congruent to others through reflections (flips) and/or rotations.
- Exchange hexominoes with another group. Check to see that none of their hexominoes are congruent. Mark any that you think are the same. Be ready to justify your findings.
- Return the hexominoes. Check yours to see if any duplicates were found.

Part 2

What if... ***Tino wants to send his friend a new basketball for his birthday? The basketball will be packed in a cubic cardboard carton that measures 12 inches on each edge. How could Tino construct the carton from the sheet of cardboard?***

- Examine your hexominoes and predict which of them could be folded along the lines to form a cubic cardboard carton.
- Build models of your selections. You may use whatever materials you have available.
- If packing tape is used to secure the unfolded edges of the carton as it is assembled, decide whether any of your models requires less tape than others. Be ready to explain your reasoning.

Write a short paragraph describing the differences between hexominoes that will form cubic cartons and those that won't. Include any diagrams that might help illustrate your points.

Gulliver's Shapes

Part 1

In Jonathan Swift's novel Gulliver's Travels, Gulliver finds himself in the great city of Lilliput. Here everything, including its citizens, the Lilliputians, are very small in comparison to Gulliver and his belongings. If the Lilliputians had a set of Pattern Blocks in their schoolroom, Gulliver would find them much too small for his use. Can you build enlargements of each of the Pattern Block shapes that Gulliver could use?

- Working with your group, try to build at least 4 different-sized enlargements of each shape in the Pattern Block set. For each enlargement, use only blocks that are congruent to the original shape, and only one layer of blocks.
- Record the number of blocks used to build each enlargement. Look for patterns as you work. If you are unable to build an enlargement of a shape, try to figure out why.
- Trace only the outlines of each enlargement onto unlined paper and cut them out.
- Compare the angle measures of each enlargement to those of the original Pattern Block shape. Compare the lengths of the sides of each enlarged shape to those of the original shape.
- Summarize the findings of your group.

Part 2

What if... *Gulliver found himself in Brobdingnag, an imaginary land of giants, where students used Pattern Block shapes that were much too large for him to work with? Can you create reductions of each of the enlarged Pattern Block shapes that Gulliver could use?*

- Exchange the cutouts of the biggest enlargement for each Pattern Block shape with those of another group. Imagine these enlargements are congruent to the Pattern Blocks used in Brobdingnag.
- Working from your cutouts, try to build at least 3 different scaled-down, similar versions of each Brobdingnag block. You may use a protractor and ruler and/or paper folding techniques on the cutouts, but you may not use Pattern Blocks or your other enlargements.
- Trace the outlines of these smaller shapes onto unlined paper.
- Compare your reductions to the original cutouts by considering corresponding angle measures and corresponding side lengths.
- Be ready to explain your methods.

Suppose a friend wants to come to your house after school to visit. Using your knowledge of similarity, draw a scale map of the route he must follow from the school to your house. Be sure to include the scale factor, landmarks, and street names.

Tan's House

Part 1

Tan is an architect who builds houses like the one shown based on Tangram shapes. He needs to construct a set of Tangram pieces whose sides measure three times the lengths of the sides of those in his original set. How might he do this?

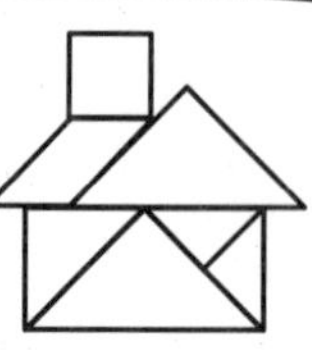

- Working with partner, use your Tangram set as a blueprint to construct a larger set of Tangram pieces. Make the ratio of the side lengths of each plastic Tangram piece to the side lengths of the corresponding larger piece 1:3.
- Copy the new pieces onto Tangram paper and cut them out. Call this new set *Tangram 2.*
- Be prepared to explain how you constructed each piece, and how you know each piece is similar to the original Tangram piece.

Part 2

What if... *Tan decides the pieces in Tangram 2 are not large enough? What if he wants a third set of Tangrams where the medium triangle from Tangram 2 becomes the small triangle in this new set?*

- Using the original plastic Tangram set and/or your new set, *Tangram 2,* as a blueprint, construct a larger set of Tangram pieces in which the medium triangle of *Tangram 2* becomes the small triangle of the third set.
- Copy the new set of pieces onto Tangram paper and cut them out. Call this new set *Tangram 3.*
- Decide whether or not each new piece is similar to the piece in the original plastic Tangram set. Justify your answer.

Using all of the pieces in the original Tangram set, design a house that is different from Tan's house. Trace its blueprint on paper, indicating where each shape is located. Then, using either *Tangram 2* or *Tangram 3,* build a similar enlargement of the original house and trace its blueprint on paper, indicating the position of its shapes.

Rooftop Triangles

Part 1

A lumber yard stocks wooden beams used to make triangular trusses for roof supports. They stock beams in lengths of 1 yard, 2 yards, 3 yards, 4 yards, and so on, up to 10 yards. What combinations of lengths can be used to make the triangular trusses?

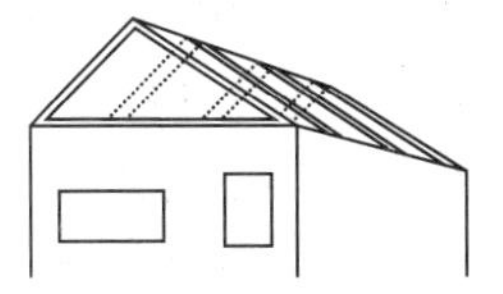

- Working with a partner, use Cuisenaire Rods to build models of triangular trusses. Arrange the rods so that each rod touches a corner of the other rods.

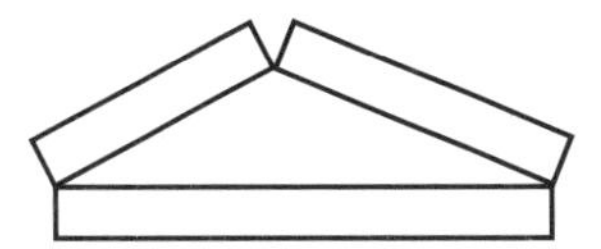

- Try at least 15 different 3-rod combinations. Record the combinations of lengths that form triangles in one list and the combinations that do not form triangles in another list.
- Organize and analyze your data. Try to figure out why some 3-rod combinations will form triangles while others will not. Be ready to explain your reasoning.

Part 2

What if... ***the lumber company wants to list in its merchandise catalog all possible triangular roof trusses it can supply from its stock? How can they organize their list to be sure to include all possible combinations?***

- Working with your partner and another pair of partners, find all possible trusses the lumber company can build. Use your Cuisenaire Rods to model the trusses and to check your combinations.
- Organize your data in a way that you think would be practical for the catalog listing.

Write a letter to the owner of the lumber yard explaining how to determine possible lengths for the third side of a triangular roof support (without physically trying them) if the lengths of two sides of the triangular support are known.

Shelf Brackets

Part 1

Kari is building shelves to hold her stereo, books, and pictures. The shelves are to be various widths, each requiring 2 identical right-triangular wooden support brackets underneath. If Kari has a thick piece of wood measuring 24 in. x 24 in., what are the different-sized right-triangular brackets she can cut from the piece of wood?

- Working with a partner, make as many different-sized right triangles as you can on your Geoboard. Use the Geoboard pegs for the vertices of your triangles.
- Imagine that your Geoboard represents Kari's 24 in. x 24 in. piece of wood and that your triangles are patterns for the shelf brackets. Find the area of each of the shelf brackets you modeled. (Hint: First figure out the dimensions and area represented by each small Geoboard square.)
- Draw each right-triangular bracket on geodot paper and record its area. Compare your drawings to make sure they are all different.
- Be ready to explain how you know you have found all possible solutions.

Part 2

What if... ***Kari wants to construct 4 shelves of varying widths? If each shelf requires a set of 2 identical right-triangular brackets with one length the width of the shelf, how can she make maximum use of the 24 in. x 24 in. piece of wood?***

- Working with your partner, make 2 copies of each right-triangular bracket pattern on geodot paper and cut them out.
- On a clean sheet of geodot paper, connect the outermost set of dots to create a square representing Kari's 24 in. x 24 in. piece of wood.
- Working with your cutouts, find a way to fit 4 pairs of different-sized triangles in the ruled-off geodot square. Start by arranging one pair of identical triangles, then a second pair, a third pair, and finally a fourth pair. The triangles may be reflected and/or rotated as they are placed, but they may not overlap. Try to minimize the amount of wood that would be wasted once the triangular supports are cut out.
- Once you've found an arrangement that you like, trace it onto the paper.
- Calculate the total area used by the 8 brackets in your arrangement, and the area of any leftover wood.
- Be ready to tell about the strategies you used for solving Kari's problem.

Write a letter to Kari explaining how she could use a Geoboard to find the areas of the triangular supports that can be cut from her piece of wood. Be sure to include instructions for triangles whose sides are not parallel to the edges of the board, as well as for those whose are. Include any diagrams that might be helpful.

Hydroponics

Part 1

Hydroponics is the science of growing plants in water containing dissolved inorganic nutrients. The plants in their water containers are supported on isosceles triangular frames as they grow. Different-sized plants require different-sized frames. Using the circular Geoboard, how many different triangular frames can you design?

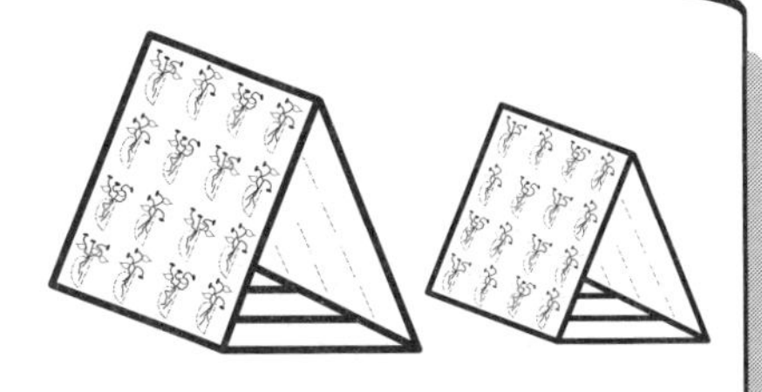

- Working with a partner, create as many different-sized isosceles triangles as you can on your circular Geoboard. Remember that isosceles triangles have at least 2 congruent sides. The vertices of your triangles can be the center peg and 2 pegs on the circle, or they can be 3 pegs on the circle.
- Find the measure of each angle of your isosceles triangles. To do this, first find the measure of the arc formed by any two consecutive pegs on the circle. (Remember that the measure of the entire circle is 360°.) Then use the following formulas to help:
    - The measure of a central angle is equal to the measure of the arc it intercepts.
    - The measure of an inscribed angle is half the measure of the arc it intercepts.

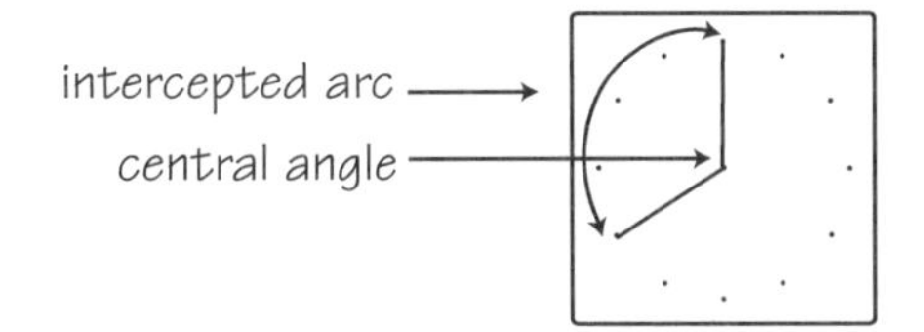

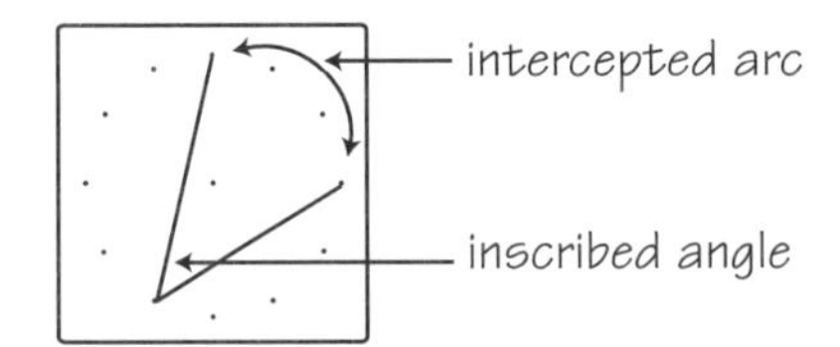

- Draw each isosceles triangle on circular geodot paper and record its angle measures.
- What observations can you make about your triangles?

Part 2

What if... ***the hydroponics frames were to be sold in sets containing one small triangular frame and one large triangular frame? How would you select the pairs of frames that would be packaged together?***

- Working with your partner, cut out each of your isosceles triangles.
- Think of a way to group your "triangular frames" into sets of two for packaging.
- Investigate the properties shared by the triangles in each of your sets. Compare angle measures, types of angles, and side lengths.

List at least five geometric properties that you learned about in this activity that you didn't know before. Use diagrams to help illustrate each property.

Sal's Similar Sails

Part 1

Sal is a sail maker at a marina in a small coastal village. He constructs sails using combinations of different-sized pieces of canvas. If Sal has 7 pieces of canvas shaped like Tangram pieces, how many different-sized sails can he make?

- Working with a partner, make a model of a triangular sail using 2 or more Tangram pieces.
- Find the number of units in the area of your model. Let the area of the smallest Tangram triangle represent 1 unit.
- Trace your model on a sheet of paper. Include the outlines of the shapes you used to build it. Record its area.
- Continue to make and record triangles until you have made models of all the different-sized triangular sails that you can. Then cut out each triangle.
- Compare the triangles and be ready to discuss their similarities and their differences.

Part 2

What if... ***Sal had to stitch roping onto each side of the sails to help them maintain their shape? How much rope would Sal need for each sail?***

- Using the length of the shorter side of the smallest Tangram triangle to represent 1 unit of length, find the side lengths of each triangle from Part 1.
- Find the perimeter of each triangle.
- For each model, record the lengths of the 3 sides and the perimeter.
- Pick any 2 of your triangles and compare them. Compare side lengths, perimeters, and areas. What relationships can you find? Test these relationships using a different pair of triangles. What generalizations can you make about the triangular sails?

Write a brief letter to Sal, listing and explaining all the different properties you discovered about the possible sails he can construct. Include any diagrams that might be helpful.

Geoboard Challenge

Part 1

Audrey has invented a Geoboard game called Geoboard Challenge. She has written 14 clues, each describing a shape that may or may not be possible to create on a Geoboard. To play the game, you must try to create the shape on the Geoboard, if it exists. Play Geoboard Challenge and see how many points you can score!

- For each description below, decide if it is possible or impossible to create the shape on your Geoboard. If you find a solution, award yourself 1 point. If you can find more shapes that fit the description, award yourself another point for each solution.
- Record your solutions on geodot paper. Label them with the number of the corresponding clue and cut them out. If you think it is impossible to make the shape described, be ready to explain why.

Geoboard Challenge

1. Find a triangle that contains 5 interior pegs.
2. Find a triangle that contains 6 interior pegs.
3. Find a triangle that contains 7 interior pegs.
4. Find an isosceles triangle that has one obtuse angle.
5. Find an isosceles triangle that has a right angle.
6. Find an isosceles triangle that has no congruent angles.
7. Find a rectangle that contains 3 interior pegs.
8. Find a rectangle that contains 5 interior pegs.
9. Find a square that has side lengths greater than 2 units but less than 3 units.
10. Find a square that has an area of 6 square units.
11. Find a hexagon that has no parallel sides.
12. Find a hexagon that has 3 parallel sides.
13. Find a hexagon that has no congruent sides.
14. Find a hexagon that has all sides congruent.

Part 2

What if... ***you wanted to make up your own version of Geoboard Challenge to play with a friend? What clues would you create?***

- Create a list of 10 clues for shapes that may or may not be possible to make on the Geoboard. Make up an answer key. Keep it hidden from your partner.
- Exchange clues with your partner.
- Take turns trying to solve a clue from each other's list. Score 1 point for each solution you can find for a given clue. Score 1 point if you correctly identify a description as impossible to satisfy and are able to explain why. Score 1 point each time you stump your partner.
- Keep track of your score and your partner's score to see who scores the most points.

Write a clue that has exactly 5 solutions. Draw your solutions on geodot paper. Then choose one of the solutions and write a clue for which the chosen solution is the only solution.

Braille Puzzles

Part 1

Michel works for a company that manufactures jigsaw puzzles for the visually impaired. His job is to write step-by-step instructions for assembling the puzzles. The instructions are then translated into Braille for the visually impaired. How accurately would you be able to write these kinds of instructions?

- Sit back-to-back with your partner (or with a barrier between the two of you) so that you will not be able to see each other's work. Make a puzzle design using at least 5 of the 7 pieces from a set of Tangrams.
- On a piece of paper, record your puzzle design, including the placement of the individual Tangram pieces used. This recording will become the answer key to your puzzle.
- On a second sheet of paper, write the step-by-step instructions for making your puzzle design. Use mathematical terms and be concise. Do not use any drawings.
- Exchange puzzle instructions with your partner. Try to follow your partner's instructions for making his or her puzzle design.
- Once you and your partner have finished the puzzles, compare the puzzles to their answer keys. Discuss how the designs compare and how the written instructions might be improved.

Part 2

What if... ***Michel is asked to write instructions for the jigsaw machine that is used to cut the frame of the finished puzzle design? His instructions will then be programmed into the machine, telling the jigsaw how to move around the border of the puzzle to make the appropriate cuts. How might he do this?***

- Sitting back-to-back with your partner, make a new puzzle design using at least 5 of the 7 pieces from a set of Tangrams.
- On a piece of paper, trace the outline, or "frame," of your puzzle design. This recording will become the answer key to your programming instructions.
- On a separate sheet of paper, write the commands you would use to program the jigsaw machine to cut the frame for your puzzle. To do this, select one vertex of your puzzle design as the starting point. Then, using a ruler and protractor as measuring tools, write a series of step-by-step commands that generate a path around the perimeter of the puzzle design, arriving back at the starting point.

- Exchange instructions with your partner. Try to follow your partner's instructions for making his or her puzzle frame.
- Once you and your partner have finished the frames, compare them to the answer keys. Discuss how the frames compare and how the written instructions might be improved.

Make a list of 10 mathematical terms that you think are helpful in describing Tangram designs. Define each of the terms, including diagrams where helpful.

Circuit Boards

Part 1

An electronics company manufactures circuit boards in convex shapes. The board must be convex so that any two points on it can be joined with a straight segment of wire lying entirely on the surface of the board. What circuit board shapes can they design using a set of Tangrams?

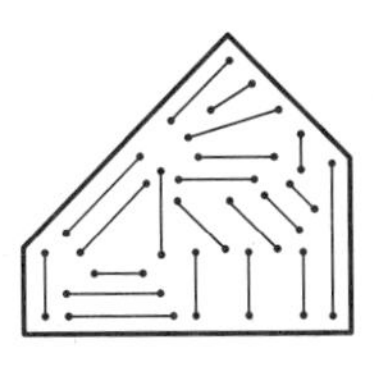

- Work with a partner. Using the complete set of 7 Tangram pieces, design as many different convex polygon shapes as you can to represent the circuit boards.
- Use a piece of uncooked spaghetti to check that your shapes are convex. Here's how: If you choose any two points on the shape and connect them using a piece of spaghetti, and the portion of the piece of spaghetti that lies between the two points remains inside the shape or lies along one of its borders, the polygon is convex. If any portion of the piece of spaghetti between the two points falls outside the shape, the polygon is concave.
- Trace around the border of each convex shape to record your circuit boards. Then cut them out and sort them according to the number of sides they have.
- Compare the cutout circuit board shapes to make sure they are all different. Be ready to discuss the shapes that you found.

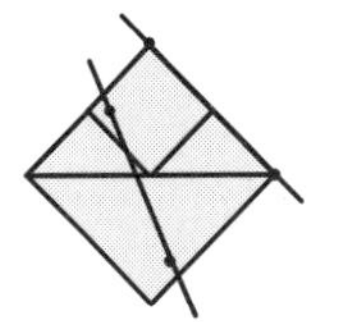

convex

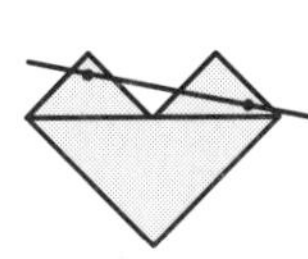

concave

Part 2

What if... *the electronics company wants to expand its production capabilities? If it now can design circuit boards based on shapes formed from 2 sets of Tangrams, what larger convex circuit boards can they produce?*

- Using 2 sets of Tangram pieces, design as many different convex polygon shapes as you can to represent the new circuit boards.
- Check using a piece of spaghetti to make sure your shapes are convex.
- Trace around the border of each convex shape to record your circuit boards. Then cut them out and sort them. Compare the cutout circuit board shapes to make sure they are all different.
- Compare your new shapes to the shapes you made in Part 1. What observations can you make?

Describe the difference between a convex polygon and a concave polygon. Then state two methods that could be used to determine whether a polygon is convex or concave.

Star Search

Part 1

Islamic art is known for its rich, intricate geometric patterns, all of which are based on the circle. In the Islamic culture, the artist and the mathematician are one and the same person. The art uses underlying square or triangular pattern grids to create mosaic designs. Some designs are based on diagonals of polygons. What can you discover about these diagonals?

- Work with a partner. Each of you should make a polygon with a different number of sides on the circular side of your Geoboard. The vertices of your polygons should be located at pegs on the circle, not at the center peg.
- Select any vertex of your polygon and make as many diagonals as possible from that vertex.
- Draw your polygon and diagonals on circular geodot paper. Record the number of sides and vertices in the polygon, the number of diagonals drawn from one vertex, and the number of triangles created in the interior.
- Repeat the process for polygons with different numbers of sides. Try at least 6 different examples. Be sure to record your polygons and data.
- Cut out your geodot drawings and organize them with your partner's drawings. Look for patterns and be ready to talk about what you noticed.

Part 2

What if... ***the Islamic artisan/mathematician wanted to use diagonals to create stars with different numbers of points? How many diagonals would be needed to create these stars in polygons with different numbers of sides?***

- From your work in Part 1, select any polygon with 5 or more sides and recreate it on your Geoboard. Make as many diagonals as you can from each vertex of the polygon.
- Draw your polygon and diagonals on circular geodot paper. Record the number of vertices in the polygon, the number of diagonals drawn from each vertex, and the total number of diagonals in the polygon.
- Repeat the process using other polygons from the first activity. Try at least 6 different examples. Be sure to record your polygons and data.
- Look for patterns in your data. Also look to see whether you think some of your stars are esthetically more pleasing than others. Be ready to talk about your observations.

Picture a 20-sided polygon. Figure out the number of diagonals that could be drawn from one of the vertices of the polygon, the number of triangles that would be formed by these diagonals, and the total number of diagonals that could be drawn in the polygon. Then write a convincing argument supporting your conclusions.

Spider Web Site

Part 1

Most children learn the song about the "itsy-bitsy spider" that goes up the water spout. Suppose the spider decided to build a web inside the water spout. If it must start by weaving a triangular web whose vertices touch the inner walls of the circular spout, can you help the "itsy-bitsy" spider figure out the number of ways it can start weaving its web?

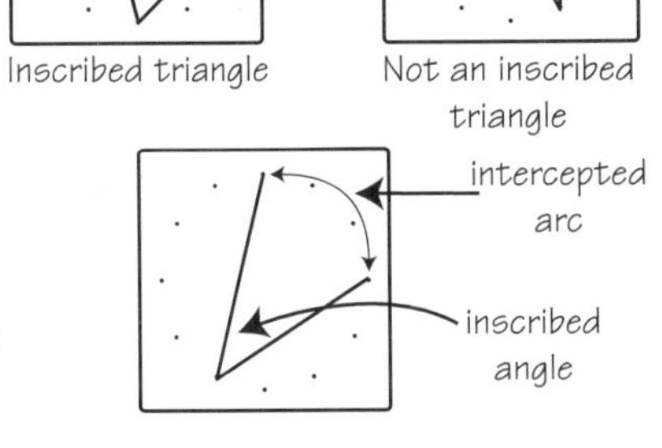

Inscribed triangle

Not an inscribed triangle

- Use a circular Geoboard to represent the inside of the water spout. Working with a partner, create an inscribed triangular "web" on your Geoboard. Remember that in inscribed triangles, all 3 vertices must lie on the circle.
- Find the measure of each angle in your triangular web. To do this, remember that:
  - A circle contains 360°.
  - The measure of an inscribed angle is half the measure of the arc it intercepts.

intercepted arc

inscribed angle

- Using a ruler, record your triangle on circular geodot paper. Label each angle with its measure.
- Find as many different inscribed triangular webs as you can and record your findings.
- Organize your work and look for patterns.

Part 2

What if... ***the "itsy-bitsy" spider decided to try different polygon shapes for the beginning of its web? How can you use what you know about inscribed angles to find the measures of the angles in its new web?***

- Decide with your partner whether the new polygon web will contain 4, 5, or 6 sides.
- Create an inscribed polygon with the agreed upon number of sides on your circular Geoboard. Figure out the number of degrees in each inscribed angle of the polygon web.
- Using a ruler, record the inscribed polygon on circular geodot paper. Label each angle with its measure.
- Find as many different inscribed webs as you can having the given number of sides. Record your findings.
- Organize your work and look for patterns.

Write an explanation of how the sum of the measures of the angles of a triangle is related to the number of degrees in a circle. Use words such as *inscribed polygon, vertex, inscribed angle, intercepted arc, angle measure,* etc. in your explanation.

Patangles

Part 1

Pat Bangle loves to build new angles (which he calls "Patangles") based on combinations of angle measures found in the 6 Pattern Block shapes. In order to create these "Patangles," Pat needs to know the measures of the angles he has to work with. Can you find these angle measures?

- Working with your partner, find the measure of each interior angle of each Pattern Block shape. To get started, remember that the green triangle has 3 congruent angles, and that the sum of the angles in any triangle is 180°.
- Trace each Pattern Block shape on a piece of white paper and record its angle measures. Also record the number of angles in the polygon.
- Using a ruler, extend one side of each polygon at each vertex. These angles formed outside the polygon are called exterior angles. Use your Pattern Blocks to find the measure of each exterior angle of each Pattern Block shape. Remember that the measure of a straight angle is 180°.

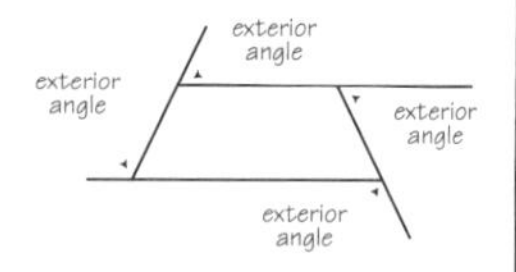

- Record the measure of each exterior angle of the Pattern Block shapes.
- Bisecting an angle means dividing it into 2 congruent angles. Use paper-folding to bisect each of the interior and exterior angles of each Pattern Block shape.
- Label each new angle with its measure. Look for angle measures that you have not already found.
- Organize your data in a chart. Look for patterns in your data.

Part 2

What if... ***Pat Bangle invited you to help him build Patangles? What Patangles with measures between 0° and 360° can you create?***

- Working with your partner, build a new angle based on combinations of the Pattern Block angle measures. Here's how:
  - First look at the interior and exterior angles you found in the first activity.
  - Use addition, subtraction, or a combination of both on these angles to create a new angle (a Patangle).
  - Your Patangle should have a measure that is different from that of any angle you have found so far.
- Trace your Patangle on a piece of white paper, showing the angles used to build it. Record its measure and the measure of the angles you used to build it.
- Find as many Patangles as you can. Record each one as described above.
- Be ready to discuss any patterns in the measures of your Patangles.

Write a letter to Pat Bangle describing the method(s) you used to create Patangles. Include any diagrams that might be helpful in illustrating your methods.

M.C. and Me

Part 1

M.C. Escher, the noted Dutch artist who lived from 1898 to 1972, often visited the Alhambra in Spain to gain inspiration for his work. Many of Escher's tessellating creatures and geo-metric shapes were based on the magnificent tile patterns he found while visiting there. How can you use Pattern Blocks to create tessellating designs?

- Working with your partner, find the measure of each angle of each Pattern Block shape. To get started, remember that the green triangle has 3 congruent angles, and that the sum of the angles in any triangle is 180°.
- Trace each Pattern Block shape on paper and record its angle measures.
- Draw a point in the center of a clean sheet of paper. Using only green triangles and placing them so that a vertex of each triangle touches your point, determine the number of triangles needed to surround the point completely.
- Trace each triangle as it appears in the tessellating pattern and determine the sum of the angle measures surrounding the point. Record this sum and the number of triangles needed to surround the point.
- Repeat the process with the other five Pattern Block shapes. For some shapes there may be more than one way to arrange them. Record your different arrangements and look for patterns in your data.
- Select your favorite tessellating arrangement, and explore ways to connect several identical arrangements together to form a larger tessellating design. Remember that in a tessellation, there can be no spaces or gaps between the shapes.

Part 2

What if... you wanted to build tessellations of the type described as "mathematical mosaics?" If a mathematical mosaic is a design that has the same arrangement of shapes surrounding every vertex in the design, what mathematical mosaics can you create using Pattern Blocks?

- Working with your partner, select 3 or more Pattern Block shapes that you might like to use in your mathematical mosaic.
- Draw a point in the middle of a clean sheet of paper and try to create an arrangement that completely surrounds your point using the shapes you selected. If you are unable to complete the tessellation around the point, change one or more of the Pattern Block shapes until you obtain a tessellation.
- Explore ways to extend your tessellation to form a larger design made from the same shapes. In order for your design to be a mathematical mosaic, the arrangement of shapes surrounding every vertex in your design must be exactly the same.
- Experiment with building mathematical mosaics using other combinations of Pattern Block shapes. See what you can discover about the shapes and designs that can be used to form mathematical mosaics.

Find five tessellating patterns in your home or environment. Describe the types of polygons used in the designs and their angle measures. Decide whether the tessellating patterns are mathematical mosaics and support your reasoning. Include a sketch of each pattern.

Modular Seating Cubes

Part 1

Elena is rearranging the furniture in her room. In addition to her bed and a dresser, she has 4 different-colored modular seating cubes. She would like to consider possible arrangements of the seating cubes without actually moving them. Can you help Elena by drawing different views of possible arrangements?

- Use 4 different-colored Snap Cubes to represent the 4 modular seating cubes. Connect the 4 Snap Cubes in any way you choose to model a potential seating arrangement for Elena's room.
- Using isometric dot paper, draw as many different views of your 4-cube arrangement as possible. Color your drawings, showing the color of each cube in your model.
- Compare your cube arrangement and isometric drawings with those of other members of your group.
- Be ready to explain how your drawings show three dimensions.

Part 2

What if... ***Elena bought a fifth seating cube for her room? Could you design a 5-cube seating arrangement and draw different views of it so that Elena could arrange the seating cubes working from your drawings?***

- Use 5 different-colored Snap Cubes to represent the 5 modular seating cubes. Connect the 5 Snap Cubes in any way you choose to model Elena's new seating arrangement. Keep your model hidden from your partner.
- Using isometric dot paper, draw at least 4 different views of your 5-cube arrangement. Color your drawings, showing the color of each cube in your model.
- Exchange your dot-paper drawings with a partner. Using the drawings, try to build each other's seating arrangements. Then draw a view of the arrangement that is different from the views your partner drew.

Write about the methods you used to draw your models on dot paper. Describe any differences that exist between the actual models and the two-dimensional representations, such as relative side lengths, angle measures, and changes in shape. You may want to use diagrams to illustrate your explanations.

Pentacube and Hexacube Twins

Part 1

A pentacube is a Snap Cube structure made from 5 cubes. A pop-up pentacube is a pentacube which, no matter how it is placed on a table, will have at least one cube that does not touch the table. How many different pop-up pentacubes can you make?

- Working with a partner, build as many pop-up pentacubes as you can. Check to make sure that each of your structures, no matter how it is positioned on your table, has at least one cube that does not touch the table.
- Compare your pentacubes and eliminate any duplicates. Rotating, flipping, and turning the structures may help you identify pentacubes that are identical.
- Pop-up pentacube twins are pentacubes that are mirror images of each other. Look at your models and see if any of them are pop-up pentacube twins. Draw each set of pentacube twins on isometric dot paper, drawing them in positions that make it easy to see that they are reflections of each other.
- Draw your other pop-up pentacubes on a separate piece of isometric dot paper. Try to figure out why these pentacubes have no twins.

Part 2

What if... ***you had a drawing of a pop-up hexacube? Could you draw and build its twin?***

- Using 6 Snap Cubes, build a pop-up hexacube. Keep your hexacube hidden from your partner.
- Draw your pop-up hexacube on isometric dot paper. Exchange drawings with your partner.
- Working from your partner's drawing, draw the mirror image (reflection) of the structure on the isometric dot paper.
- Now build the two hexacube structures (the original and its reflection). Examine them from all perspectives to make sure that they are hexacube twins. Then check your hexacubes with the original structure made by your partner.

Write about the differences and similarities between a reflection and a reproduction. You may find it helpful to use a photograph of yourself and a mirror to make the comparisons. Include any diagrams that might help illustrate your points.

Slice 'n' Dice Cubes

Part 1

Laura constructed some cube structures using Snap Cubes. These structures, which she called Slice 'n' Dice Cubes, measured 2x2x2, 3x3x3, and 4x4x4. Once the Slice 'n' Dice Cubes were assembled, they were dipped in red paint and allowed to dry. When Laura took these cubes apart, she was surprised to find that there were relationships between the dimensions of the cube structures and the number of cubes with paint on exactly 0, 1, 2, or 3 faces. What did she discover?

- Working with a partner, build your own Slice 'n' Dice Cubes measuring 2x2x2, 3x3x3, and 4x4x4. Record the number of Snap Cubes used to build each cube structure, and the number of its vertices (corners), edges, and faces.
- Imagine dipping your Slice 'n' Dice cubes in red paint. (Optional: Use stick-on dots to mark the painted faces of each Snap Cube.)
- For each structure, determine the number of individual Snap Cubes that have paint on exactly 3 faces, on exactly 2 faces, and on exactly 1 face. Determine the number of cubes with no painted faces.
- Organize and record your data. Look for patterns.
- Take apart the three cube structures. Build a new structure whose dimensions are 5x5x5. Imagine dipping it in red paint. Investigate this structure as you did the others and record your data.
- Look to see if you can identify any relationships between the dimensions, the total number of cubes, the number of different types of painted cubes, and the number of unpainted cubes. Be ready to explain your findings.

Part 2

What if... ***Laura decided to build different-shaped rectangular prisms using Snap Cubes? Would she find the same relationships as those she discovered for the Slice 'n' Dice Cubes?***

- Working with your partner, build at least 3 rectangular prisms using a different number of Snap Cubes for each one. Record the dimensions of each prism and the number of its vertices, edges, and faces.
- Imagine each structure being dipped in red paint. (Optional: Use stick-on dots to mark the painted faces of each Snap Cube.)
- For each prism, determine the number of individual Snap Cubes that have paint on exactly 3 faces, on exactly 2 faces, and on exactly 1 face. Determine the number of cubes with no painted faces.
- Organize and record your data. Look for patterns.
- Look to see if you can find any relationships between the dimensions of the prisms and the number of cubes with paint on a given number of faces. How do these relationships compare to those you found for the Slice 'n' Dice Cubes? Be ready to explain your findings.

Write a letter to Laura explaining how to figure out the number of Snap Cubes with paint on at least 2 painted faces in a rectangular prism whose dimensions are 4x6x11.

Saving Paper

Part 1

Juan works in the shipping department of a company that manufactures coffee mugs, each of which is packaged in its own cubic cardboard box. Juan is filling an order for 4 coffee mugs. He needs to determine how to group the 4 boxes together as a single unit for shipping and cut one piece of wrapping paper that can be used to cover the 4 boxes. How might he do this?

- Use 4 Snap Cubes to represent the 4 boxes containing the coffee mugs. Working with your partner, decide on 2 different ways to group the boxes together for shipping.
- Using Snap Cube grid paper, design 2 nets for each of your structures. Think of your nets as the wrapping paper that could be used to cover the structures without any overlaps.
- Cut out each net to see if it works. Fold on the grid paper lines and tape the common edges together. If the net doesn't work, revise the design until it does.
- Determine the number of vertices, edges, and faces on each of your "wrapped packages." Record your results.
- Make a clean copy of each net to share with the class. Be ready to show how you figured out the design for your nets.

Part 2

What if... *Juan has to fill 4 of these orders for sets of 4 coffee mugs? How might he arrange 4 copies of the net on a rectangular sheet of wrapping paper if he wants to minimize the amount of wasted paper?*

- Choose one of your 4-cube structures to represent your package. Compare all of the nets for the structure. Choose one of the nets (different from the one chosen by your partner) and make 4 copies of it.
- Using one sheet of Snap Cube grid paper, find a way to enclose the four nets in a rectangular area. Try to arrange your nets in such a way as to minimize the amount of wasted paper.
- Find the area of the rectangle enclosing the 4 nets and the area of the wasted paper.
- Compare your work with your partner's to determine which net and which arrangement saves the most paper. If you think a different net or arrangement might be more efficient, try it and see.

Write about the methods you would use to construct a net for any 3-dimensional structure. Include any diagrams that might help illustrate your methods.

1-inch Grid Paper

Tangram Paper

Geodot Paper - 4 Grids

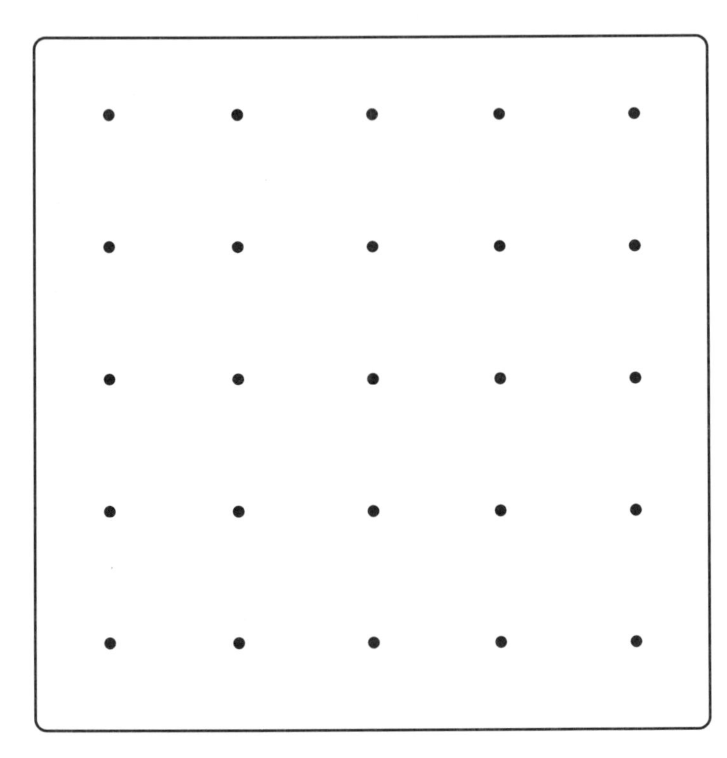

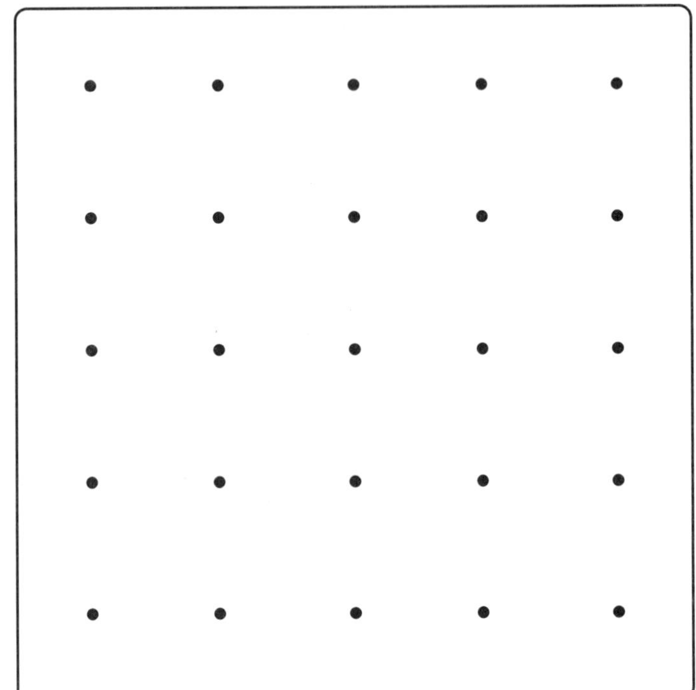

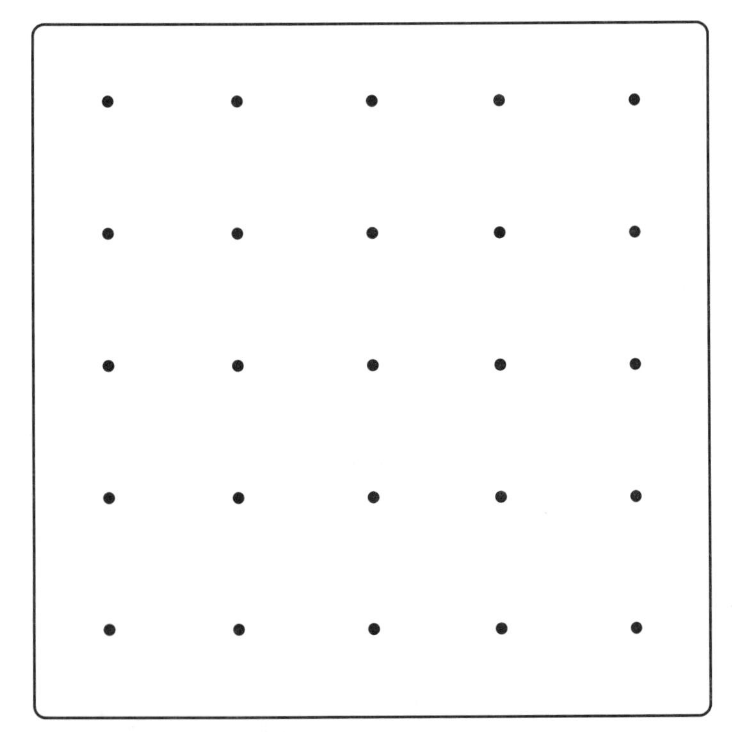

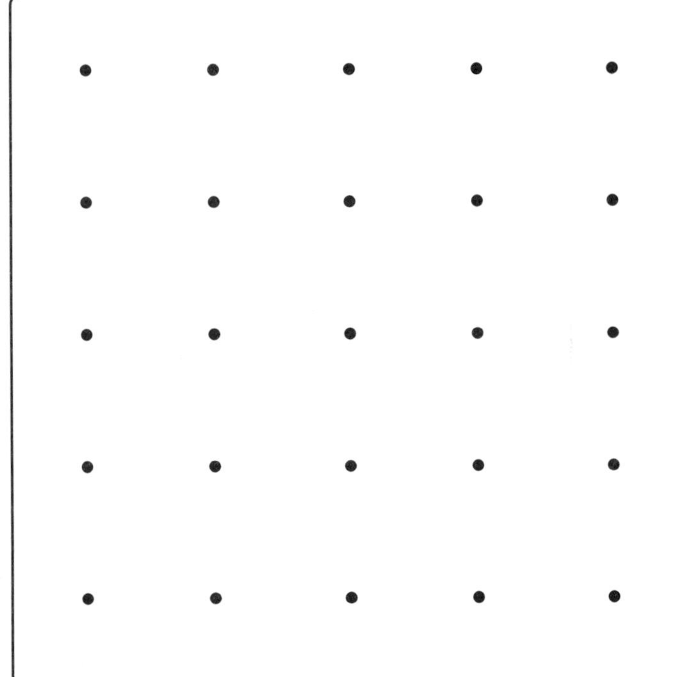

Geodot Paper - 9 Grids

Circular Geodot Paper

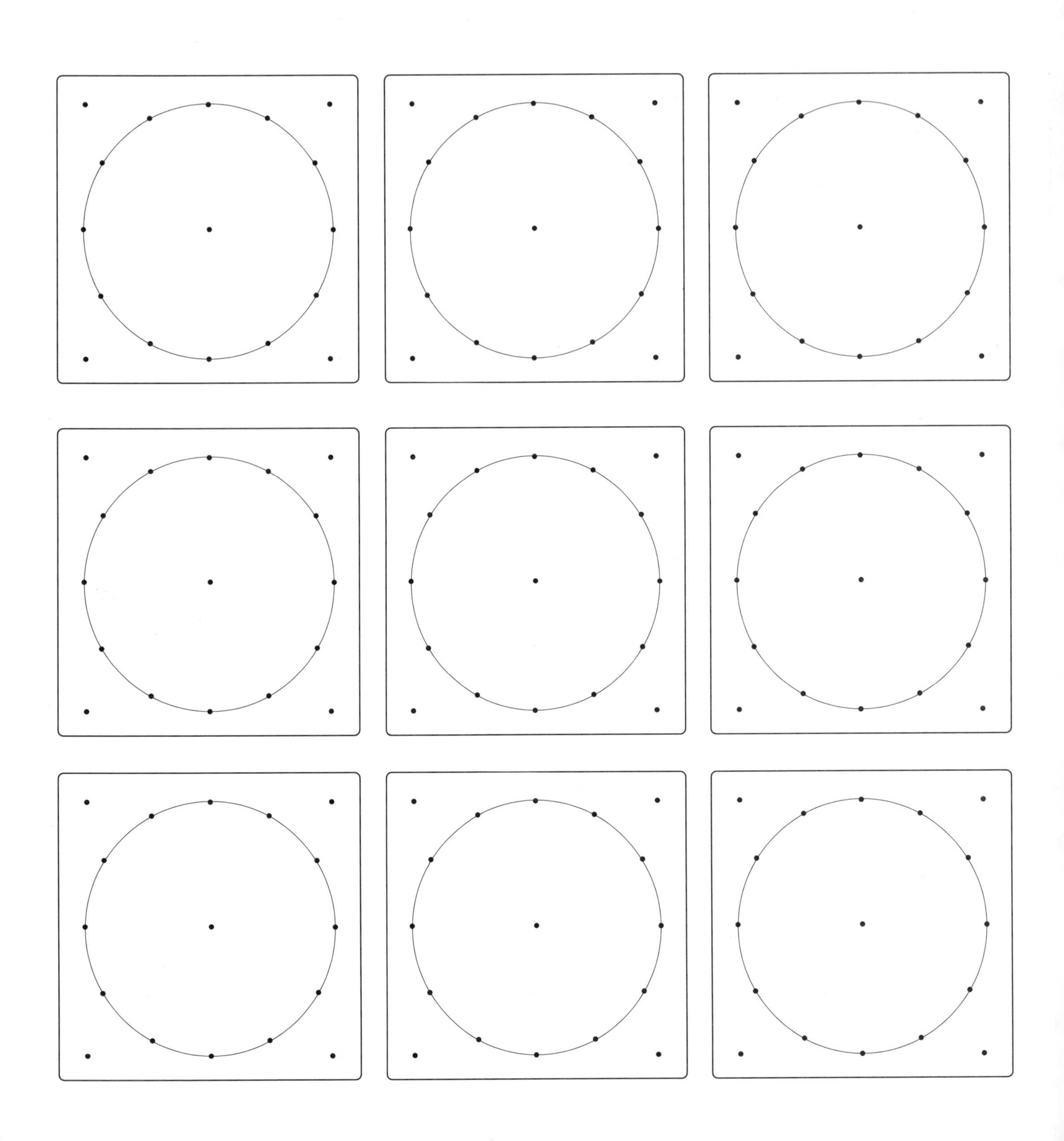

Isometric Dot Paper

Snap Cube Grid Paper